Resin Transfer Moulding

Resin Transfer Moulding

Kevin Potter
Department of Aerospace Engineering
University of Bristol
UK

CHAPMAN & HALL
London · Weinheim · New York · Tokyo · Melbourne · Madras

Published by Chapman & Hall, 2–6 Boundary Row, London SE1 8HN, UK

Chapman & Hall, 2–6 Boundary Row, London SE1 8HN, UK

Chapman & Hall GmbH, Pappelallee 3, 69469 Weinheim, Germany

Chapman & Hall USA, 115 Fifth Avenue, New York, NY 10003, USA

Chapman & Hall Japan, ITP-Japan, Kyowa Building, 3F, 2-2-1 Hirakawacho, Chiyoda-ku, Tokyo 102, Japan

Chapman & Hall Australia, 102 Dodds Street, South Melbourne, Victoria 3205, Australia

Chapman & Hall India, R. Seshadri, 32 Second Main Road, CIT East, Madras 600 035, India

---

First edition 1997

Softcover reprint of the hardcover 1st edition 1997

Typeset in 10/12 Times by Florencetype Ltd, Stoodleigh, Devon

ISBN-13: 978-94-010-6497-2 e-ISBN-13: 978-94-009-0021-9
DOI: 10.1007/978-94-009-0021-9

A catalogue record for this book is available from the British Library
Library of Congress Catalog Card Number: 96-72035

∞ Printed on permanent acid-free text paper, manufactured in accordance with ANSI/NISO Z39.48-1992 and ANSI/NISO Z39.48-1984 (Permanence of Paper).

Contents

Preface

The science and technology of composite materials has generated a large number of processes by which components can be manufactured. These range from the contact moulding approach of rolling resin into the reinforcement on simple tools to the use of capital-intensive automatic tow placement machinery. All of these processes have a single aim in common; the cost-effective meeting of the design requirements for the part to be made. To meet this aim the strengths and weaknesses of the manufacturing route must be reflected accurately in the design of the part.

There is a tendency to treat design and manufacture as two boxes which, while they overlap, can be handled separately. For processes that have a large experience base that can be called upon there may be some justification for this belief, although maximum effectiveness will always arise from a concerted design and manufacture approach. For emerging processes there is a danger that carrying over design practices from other processes will be, at best, non-optimal.

In order to utilize a concerted design and manufacture approach it is essential that access is available to sources of information to guide both design and manufacture. In the case of resin transfer moulding (RTM) a great deal of information is available in the academic literature. Perhaps the majority of this information is concerned with details of the scientific underpinnings of the process and less is available on the apparently more mundane aspects. It is often these less immediately exciting issues that have the greatest influence on matters such as detailed design and, vitally, production costs.

This work aims to provide an adequate understanding of the basic science of RTM and to provide much needed information on the technological aspects of the process. The aim throughout is to equip both designers of components and those entrusted with their manufacture with the tools to make the best of the opportunities that RTM presents to them.

Lastly, the word **advanced** will be found throughout the text. This is a difficult word as it seems to mean very different things to different people.

In this work it can be taken as a shorthand way of saying any or all of the following: highly loaded; complex in geometry; intended for use in safety critical applications; of high moulded quality and free of defects; optimally cost-effective and so on.

Introduction

It must be stressed at this early point that RTM is not a single manufacturing process that can be dealt with in a monolithic manner. RTM is better thought of as a philosophy of manufacturing in which the resin and fibres are held apart until the last possible moment. In this it can be contrasted with those manufacturing methods where the resin and fibre are combined prior to use.

In many ways the development of an aerospace or, more broadly, advanced composites industry was permitted by the development of preimpregnated reinforcements (prepreg). This development permitted real structures and components to be designed and manufactured that could reflect the properties of high-performance fibres. A variety of individual processes, such as autoclave moulding, vacuum bag moulding, compression moulding, expanding bladder moulding, and silicone rubber expansion moulding were developed that utilized the new form of semi-finished material known as prepreg. As people became comfortable and experienced with the new material form, a design and manufacturing database grew up such that the strengths and limitations of the materials and processes were reflected in design philosophies and detailed designs. These design philosophies and detailed design features have become the norm for advanced composite products. They largely reflect the capabilities of the dominant aerospace manufacturing route, autoclave moulding.

Despite the similarities between the various processes it is unusual to refer to prepreg moulding. Each process has its own literature and the commonalities between them are sometimes lost. By contrast, RTM has as many different processes under the RTM umbrella as there are prepreg processes; but it is more or less commonplace to speak of RTM rather than, for example, rigid tool RTM with semi-rigid preforms.

All the process variants have common features. Unresinated fibres are held within a tool cavity and a differential pressure is applied to a supply of resin such that the resin flows into the reinforcement completely wetting it out. The tool may be essentially rigid, semi-rigid, or contain flexible elements. Any consolidation pressure required to give the required reinforcement volume fraction (Vf%) may be applied by mechanical clamps, from a tooling frame or press, or by the use of an internal vacuum or external applied pressure in non-rigid tooling. Reinforcements may be of any fibre, and the use of all forms has been reported, from unidirectional (UD) through woven or knitted cloths to needled and random mats and fully three-dimensional reinforcement preforms. Volume fractions from below 20% to above 60% have been reported. The reinforcement may be laid onto the mould by hand, formed to shape by the mould closure, assembled by a wide variety of preforming techniques or may utilize specially woven or braided constructions. The resin can be of a very wide range of chemistries and formulations, so long as the basic process requirements are met. Cure times can be from a few minutes to many hours. Resin injection machines can be of a very wide variety of types and production line design can be just as varied.

The focus here will be on those RTM techniques that are intended to produce components to high-quality standards for structural applications. Thus material combinations such as random glass mats and polyester resins will not be discussed in any depth. Concentration will be on the materials that can produce advanced structural components and the processes for their conversion into such products.

The age of RTM as a manufacturing process is, despite its apparently recent origins, much greater than that of any prepreg-based system. RTM can be traced back to the Marco process of the 1930s,[1] and in the 1960s work was done on the pressure injection of a high-performance matrix into an organized fibrous preform.[2] The fact that the matrix was aluminium does not detract from the fact that the process was clearly a variant of the RTM methodology.

The use of an RTM approach for the manufacture of advanced polymer matrix composites is more recent. Even so RTM was used to manufacture radomes in high- and low-temperature matrices as early as the mid 1970s.[3] Later in the 1970s RTM was used for other components such as aeroengine compressor blades.[4] Most of these early applications were driven by the need for high levels of geometrical accuracy and this is still a major driving force behind many RTM component developments. By 1980 many groups were attempting to devise manufacturing methods that could step beyond the cost and geometrical complexity limitations imposed by the baseline aerospace manufacturing processes.

At that time RTM was fairly well developed as a niche process in the general engineering composites area, and some of the early advanced RTM work was carried out at the top end of the general products area rather than in aerospace. This sort of work is exemplified by the development by British Petroleum of high-speed flywheel system components by RTM.[5, 6] The materials used were glass fibre cloths and polyester resins, but the geometrical accuracy and mechanical reliability requirements were very high, while costs had to be constrained for the proposed transport application. Many of the approaches to tooling and preforming that are in use today can be traced back to such early work in the field, and this work established the suite of advantages that RTM can bring to the design and manufacturing processes. This period also saw the introduction of aircraft propeller blades manufactured by RTM.[7]

From the mid 1980s interest in advanced RTM began to pick up. A search of one of the major databases could only find two references to **Advanced RTM** prior to 1986, with references increasing very rapidly after 1987. In this period there was much more interest from manufacturers of aerospace components, leading to programmes resulting in series production of flight hardware outside the previous niche markets by the end of the 1980s.[8]

Interest in advanced RTM has been steadily building for more than a decade. In this period reinforcements, matrices, preform techniques, injection equipment, process and flow models have all been subject to great improvements. One of my first involvements with RTM was in 1981. A high-temperature demonstrator component of complex geometry was made, but the only bismaleimide (BMI) matrix available for RTM work had all the toughness of shellac. Enormous efforts had to be put in to the design of both component and reinforcements to overcome this fundamental weakness of the matrix. Today, while BMIs are still somewhat lacking in toughness compared to epoxies for RTM, the same component would be very much easier to deal with.

Currently many applications of the technology are in operation or development and both resins and reinforcement forms specifically tailored for RTM are widely available. This upsurge in developments is directly attributable to the advantages that RTM can bring and these are outlined below.

ADVANTAGES OF RTM

1. For rigid tool RTM all dimensions including part thickness are directly controlled by the tool cavity. Surface finish replicates that of the tool, generally a smooth finish is chosen for advanced work, but matt or decorative finishes could be utilized.
2. Net shape parts can be produced, eliminating some finishing operations.
3. Many reinforcement types, such as thick or 3D wovens, stitched assemblies and braids, are difficult to mould by conventional means. All of these forms can be handled via RTM and no problems have been reported with any specific fibre types.

4. A wide variety of resin systems can be utilized. Much epoxy resin formulation for prepreg is related to flow control during consolidation, this is not required for RTM resins although the need for a low viscosity can be difficult to reconcile with toughness requirements. Resins that cure by condensation reactions or contain volatiles are not ideal. Even so, good results have been reported with acid catalysed phenolics.[9]
5. As noted above, the prepreg process stages relating to flow control and consolidation in autoclaves are not required in RTM. This can lead to simplicity in cure scheduling, faster heat-up rates for tools that are not injected at the cure temperature and generally leads to shorter overall cure cycles.
6. Because prepreg is not used the shelf-life and refrigerated storage costs associated with the use of prepreg are avoided. The use of unimpregnated reinforcements can also lead to cost savings as the cost of the prepregging itself is avoided.
7. For fixed cavity tooling, fibre volume fractions can be very well controlled, leading to very consistent mechanical properties.
8. The factors leading to porosity and voidage in RTM are somewhat different to those in prepreg moulding. With correct mould design and good process control very low or zero voidage levels are routinely achieved.
9. Experience with operating production lines has shown that defect rates in RTM production of aerospace parts can be lower than those experienced in autoclave moulding production lines. While the positioning of quality control inspection points may be different for RTM and autoclave work, good control can be imposed on RTM-based production lines. For additional security and quality control, samples of both laminate and neat resin can be obtained within an RTM mould.
10. Very complex components can be produced via RTM. Many components have shown high levels of parts integration, leading directly to cost savings. Some of the usual geometrical limitations on autoclave moulding, such as the use of bend radii several times the laminate thickness, can be eliminated through the use of RTM. Bend radii down to half the laminate thickness have been reported without evidence of interlaminar cracking.[10]

It is very difficult to make an assessment of which of these factors is the most important. They can all make a contribution to minimizing costs, but the exact mix would depend on the specific component being considered.

REFERENCES

1. Mountfield, J. (1969) Forming processes for glass fibre and resin – other methods, *Composites* **1**, 41–9.
2. Cooper, G. (1970) Forming processes for metal matrix composites, *Composites* **1**, 153–9.
3. Cray, M. (1980) Development of a polyimide resin injection process for advanced composite structures, *Proc. Symposium: Fabrication Techniques for Advanced Reinforced Plastics*, April, Salford: IPC Science and Technology, 35–9.
4. Jones, W. and Johnson, J. (1980) A resin injection technique for the fabrication of aero-engine composite components, *Proc. Symposium: Fabrication Techniques for Advanced Reinforced Plastics*, April, Salford: IPC Science and Tcchnology, 40–7.
5. Potter, K. D. (1986) The development of a GRP casing component and its manufacture by the resin injection moulding process, *Proc. 15th Reinforced Plastics Congress* British Plastics Federation, 123–5.
6. Medlicott, P. A. C. and Potter, K. D. (1986) The development of a composite flywheel rotor for vehicle applications – a study of the interactions between design, materials and fabrication. In K. Brunsch, H. D. Golden and C. M. Heckert (eds), *High Tech – The Way into the Nineties*, Amsterdam: Elsevier Science Publishers, 29–42.
7. McCarthy, R., Haines, G. and Newley, R. (1994) Polymer composite applications to aerospace equipment, *Composites Manufacturing* **5**(2), 83–93.
8. Morgan, D. (1989) Design of an aero-engine thrust reverser blocker door, *Proc. 34th International SAMPE Symposium*, 2358–64.
9. Forsdyke, K. (1984) Phenolic resins for fire and high temperature applications, *Proc. 2nd International Conference. Fibre Reinforced Composites 84*, Plastics and Rubber Institute, Paper 1.
10. Potter, K. D. and Robertson, F. C. (1987) Bismaleimide formulations for resin transfer moulding, *32nd International SAMPE Symposium*, April.

RTM theory

1

It is probably the case that in all the technical literature on RTM more pages are taken up by attempts to understand the processes that occur during the injection of resin than by any other single area of study. There are essentially two reasons for this. Firstly, the subject of how resin moves through a tool and wets out the reinforcement is central to questions of quality and production rate. Secondly, finding reliable answers to even the most basic questions has proven to be very difficult and this has acted as a spur to the further development of models.

Before going on to discuss some of the issues relating to resin flow in RTM and other aspects of RTM theory I need to make a confession. Having been using RTM processes for very many years, and having had involvement with the fabrication of more than 50 components, I have never used any flow simulation more complex than graphing out on a sheet of paper where the flow lines are likely to fall. In large part this may have something to do with the fact that none of the current models was available a decade or more ago when I first became involved with the process. On the other hand none of the failures among these component fabrication exercises, and they were by no means all successes, could be put down to a lack of understanding of the resin flow phenomena that could have been avoided by using the flow models then available. There have been a few components where the resin flow followed unpredictable paths, but the solution of these flow problems owed nothing directly to the models. Indeed, the improved intellectual understanding of difficult flow situations gained by the practical solution of these complex flow cases was used to refine models rather than vice versa.

The above refers, of course, to one investigator's experience with high-performance RTM of complex components using primarily woven cloth reinforcements and epoxy resins and requires some caveats. Firstly flow modelling and the associated cure modelling will be of critical importance in high-speed RTM and of increasing importance if injection and cure stages overlap. For most advanced RTM (as defined here) injection and cure are clearly separated in time and the proportion of the total manufacturing cycle time associated with the injection process can be rather small. Thus the minimization of injection time has relatively little impact on total cycle economics for most advanced RTM parts. Equally, it is usual to utilize high levels of in-mould vacuum for advanced RTM work and this acts to minimize problems with irregular flow.

Notwithstanding the above comments it is necessary to have a grasp of how resin moves through a reinforcement, the processes involved, and those factors that influence both flow rate and quality. In the context of complex advanced RTM components, moulding quality is usually more critical than mould fill time and although there is a common theoretical framework for mould filling and quality it may be better to put most emphasis on the quality issues. Many of these quality issues have their roots in the most basic understanding of the flow process rather than the fine details, so this understanding will be the starting point.

1.1 THE BASICS OF FLOW IN RTM PROCESSES

If we take a tool filled with a fibrous reinforcement and inject resin under pressure a bulk flow front of resin will pass through the reinforcement. In addition each bundle of fibres must be wet out by the resin, down to the level of each individual fibre. For most advanced RTM these two processes go on in parallel, individual fibres are wet out within a few mm of the bulk flow front. If we wish to gain an insight into the mechanisms of defect generation in RTM it is important to understand that there are indeed two processes operating and that they have different characteristics.

To understand the reason for this it is necessary to look at the structure of the reinforcements that are used for advanced RTM; for example, woven and knitted cloths, braids and 3D preforms. These reinforcements can be seen to consist of tows of fibre having a high fibre content, separated by regions of low or zero fibre content, with the exact architecture dependent on the reinforcement manufacturing route. When a pressurized resin is introduced to the reinforcement it flows primarily in the gaps between the tows, essentially through 'pipes' in the reinforcement.[1] The bulk flow front that this resin transport produces is followed closely by lateral resin flow through the 'leaky walls' of the 'pipes'. Whereas the bulk flow front is dominated by factors such as applied pressure, resin viscosity and the dimensions of the gaps between the tows, the lateral flow is affected largely by capillary flow and wetting behaviour. If tools are not evacuated any point-to-point inhomogeneities in the driving forces for, or resistance to, wetting flow can lead to the entrapment of small air pockets. With time and continuing flow there is a tendency for the entrapped air to find its way back into the inter-tow gaps (under the influence of the local pressure gradient) and be washed from the system; and/or to be dissolved in the resin as the local resin pressure increases. Thus voidage tends to be concentrated at out gates in mouldings that have seen insufficient resin flushing.[2] Where sized or bound fibres are used the dissolution of the binder in the resin may also be important. This tends to be the case with glass mat reinforcements, but not with the reinforcement styles more usually used in advanced work. Preforming binders tend to have less effect on this fine scale wetting as they are concentrated between the tows.

Three observations can be used to support this view of the fundamental processes involved in RTM.

- Measurements of flow rate at a constant pressure and viscosity give very different results if the measurement is carried out on spreading flows rather than flow through a fully wet out bed of fibres.[3]
- If a dyed resin is injected into a bed of glasscloth that has already been wet out with undyed resin the dyed resin can be clearly seen to fill the gaps between the tows and not to displace previously injected resin to any great extent. Equally, if air is injected into tools late in the injection cycle it tends to remain in the inter tow gaps and can be purged out by additional resin flows (my own observations in unpublished work).
- If a resin is injected that does not naturally wet out the fibres, the resulting laminates have a high level of porosity trapped within the tows (my own observations in unpublished work).

Simply put, the bulk flow rate is directly proportional to the applied pressure gradient and inversely proportional to the resin viscosity and length of the fibre bed that has already been impregnated. There is some evidence that a simple measure of viscosity is not sufficient and that non-Newtonian behaviour is important (especially at high flow rates such as might be seen in structural reaction injection moulding),[4] but the simple picture is adequate here. The constant of proportionality in the equation is known as the permeability. This is related, although not in any simple way, to the porosity of the bed $(1 - Vf)$. The lack of simplicity arises because, as noted above, the geometry of the reinforcement, the size of tows, their position in space and the gaps between them, are critical. Permeability is a strongly negative function of Vf%, whatever the internal structure of the reinforcement. For an arbitrary reinforcement the permeabilities in the X, Y and Z directions will be different. Even for balanced 0.90 woven cloths the Z direction permeability would be expected to be different to that in X or Y directions.

Currently it appears not to be possible to predict permeability directly from a knowledge of the reinforcement structure, although the methods

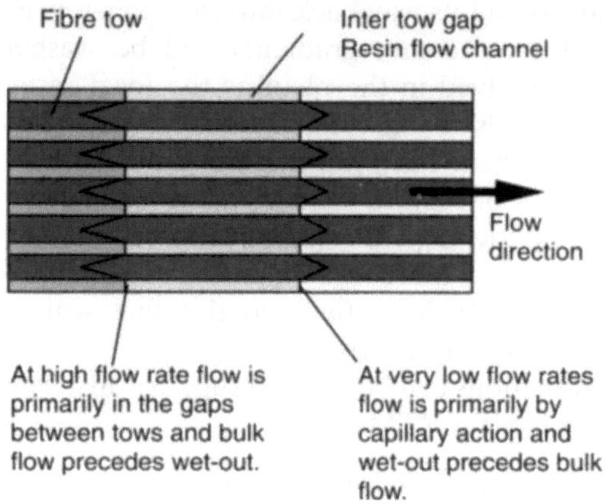

Figure 1.1 For reinforcements in which the fibres are arrayed in a regular pattern of tows and gaps between them the relationship between bulk flow front position and wet-out position is a function of the flow rate.

by which permeability can be increased are well known and have led to specific reinforcements being designed to increase flow rate.[5] On the other hand it is not too difficult to measure permeability directly, so long as the difference between spreading flows and fully wet-out flows is understood and the flow cell used is sufficiently stiff that the action of the flowing resin does not disturb the cavity dimensions.

These basic observations and the results of experiments with glass-topped moulds can be used to describe how the RTM injection proceeds in constant applied pressure cases. Initially, resin flow is rapid. Full wet-out of tows may take place a few mm behind the position of the main flow front. As noted above, once a tow has been wet out there is relatively little mixing with newly injected resin. As injection continues the bulk flow rate declines. The speed at which tows are wet out by capillary forces is much less sensitive to applied pressure, so the position of full wet-out moves closer to the bulk flow front position. If the bulk flow becomes very slow then tow wet-out can occur ahead of the bulk flow front (see Figure 1.1). If injections are prematurely halted, or a pocket of air is entrapped, resin tends to continue to flow from the inter-tow gaps into the tows themselves, producing inter-tow voidage.

The above comments assume that the fibre bed is uniformly packed (neglecting the inhomogeneities within the reinforcement). If tooling is inadequately stiff it can bow under a combination of fibre compaction and resin injection pressures. This will lead to lower fibre Vf% and much higher permeabilities. When reviewing permeability data from literature sources it is necessary to consider the design of the permeability measuring rig to ensure that it is adequately stiff to yield valid results. The author's own experience is that thick glass and a lot of steel reinforcement are required for results to be within the volume fraction tolerance ranges required; which equated to a total thickness tolerance of ±0.05 mm on the author's rig. Any results obtained from unstiffened thin glass or perspex permeability cells have to be treated with a great deal of caution. Other work has used heavy presses to overcome this source of inaccuracy, at the cost of losing a directly observable setup.[6] Another source of possible error in permeability measurements, or problems in real mouldings, is that any gaps between the edges of tools and the fibre pack act as gross channels for resin flow. Even small gaps can lead to large distortions of the flow front. For this reason, many experimental studies on RTM have used centre-ported tools to avoid edge effects.[7] For component tools, any loose blocks can act to channel resin, perhaps in unwanted ways. The intent is to fill every gap between fibres of a few microns diameter, it is hardly surprising then that the resin will penetrate to every corner of the tool.

In cases where the flow rate is constant (i.e. constant volume supply, rather than constant pressure) the position of the flow front relative to the point of full wet-out will only change if the shape of the flow front changes during injection. Thus for a parallel flow the relative positions will be constant, assuming no change in process conditions. For an expanding flow front, such as seen in a centre ported tool, the relative positions of the flow and wet-out fronts will still vary throughout the injection. For flow rate to be held constant requires that the applied pressure be constantly increased to offset the increasing viscous drag; this can lead to very high pressures being applied to the tool and other system elements. Most flow modelling work seems to have been carried out assuming constant pressure injection. The bulk of RTM production work probably uses constant flow rate machinery, although the use of constant pressure resin supplies is more common in advanced work.

A more complete review of RTM flow theory is given in section 1.2. below. This is not intended to be in any way exhaustive and may safely be skipped through by those whose interest is primarily technological.

1.2 RTM THEORY

1.2.1 Introduction

A full theoretical treatment of the RTM resin injection and cure processes would have to include many factors. To model fully the RTM moulding process necessitates the inclusion of other factors, and the modelling of

RTM as a manufacturing process requires still more factors to be included. The notes here refer exclusively to the resin injection and cure processes. Other factors are taken up in sections on reinforcement manipulation, tooling, production engineering and costing.

Factors in the modelling of resin injection and cure processes

- calculation of the flow front shape, speed and position with time for an arbitrary geometry, arbitrary gating and arbitrary Vf, both locally and globally;
- calculation of tow and fibre wet-out rates;
- calculations of cure time, demould time, peak mould and component temperatures at any and all positions in the tool and component;
- calculation of resin expansion and shrinkage effects during heat up to cure and cure. (These effects have been seen to influence moulded quality in some systems.)

In order to make these calculations various input data is required. This includes:

- tool geometry and gating geometry;
- resin input pressure or flow rate profile with time;
- resin viscosity profile with time, temperature, any binder dissolution effects, and changes in shear rate during injection;
- tooling heat-up rates and thermal transfer into the resin if injection and cure take place at different temperatures or the resin is injected at a different temperature to that of the tool;
- point-to-point reinforcement permeability across the tool, in three orthogonal directions, taking into account tool stiffness and its interaction with compaction and instantaneous injection pressure, any local Vf variation due to reinforcement preforming or layup, any high-permeability pathways due to inaccuracies of fit between reinforcement pack and tool;
- surface tension and wetting angle between fibres and resin and any time, temperature, binder dissolution or cure state changes to these;
- cure rates, exothermic heat generation, thermal conductivity of wet-out reinforcements and changes during cure;
- resin volume versus temperature and cure state.

Ideally the permeability should be derived from a knowledge of the reinforcement Vf and geometry so as to produce a comprehensive theoretical base.

It can be seen from the above listings that to generate a complete theory of the RTM injection and cure processes is a considerable task. Equally, the routine measurement of the required factors for specific components would be time-consuming and I am not aware of much work on measurements of resin volume changes during heat up and cure.[8] Lastly, for practical injection and moulding cycle derivation the variabilities in all of the above inputs should be known so that robust cycles can be determined.

The most important question is: which factors are critical and at what point do simple models need to be replaced with more complex ones?

There are two reasons for carrying out process modelling.

Firstly to aid designers of components and the associated tooling to make the correct decisions as to issues such as gating options. Also to assist those responsible for production line operations to maximize profitability (not necessarily the same as throughput) and quality.

Secondly to aid in the design of new reinforcement forms and new resin formulations that are optimized to the needs of the process and its applications. Also to assist those responsible for the design of production equipment.

These two areas have rather different sets of priorities and often very different timescales. Component and tool designers usually need a very rapid response and may have to consider a wide variety of options in early phases of design; for example in terms of in and out gate positions or the balance between complex gating options and the use of high levels of in-mould vacuum. Designers most often require easy to use and simple models that adequately represent reality and are consistent with the other tools that they habitually use. Once design decisions are made there tends to be more time available for consideration of such factors as cure scheduling. It is worth noting here that a cure model that could predict peak exotherm, cure and demould times from a knowledge of tool heating rates and temperatures would only be one input into the decision as to what mould cycle to use (assuming that the injection and cure temperatures are different). Other issues would be at least as important; these might include: tool scheduling for efficient use of manpower; baseline power availability and peak power demand and the varying costs of supplying different peak power demands; capital costs, longevity, maintainability, weight, ease of incorporation into production engineering schemes and lead-times of different sorts of tooling with different heating rate capabilities; tool lifetime versus peak exotherm etc. Equally, the identification via a flow model of ideal in and out gate positions in a complex component would have to be weighed against the costs of achieving those gating positions if this led to more complex or difficult-to-use mould tools. These issues fall into the category of modelling RTM as a moulding or manufacturing process and have to date received much less attention than the injection and cure issues; although it can be argued that tools are available to production engineers in some of these areas. The critical thing to appreciate is that even if a perfect knowledge of the flow and cure behaviours could be achieved, and models produced that could give highly accurate results for the expenditure of little time or effort, the use of such models would only provide guidance to tool and component designers.

Materials designers generally have longer timescales available to them and may be in a better position to utilize more complex models. On the other hand, to gain commercial advantage new material forms will require substantial advantages over the current competition and coarser models may be able to spot these trends.

With the above caveats in mind the current state of RTM injection and cure modelling is sketched out below.

1.2.2 Gross resin flow in RTM

The usual starting point is Darcy's Law.[9] This was originally developed as a tool for the oil industry and related to how oil permeated through rocks.

Darcy' law is $V = K/n(dp/dx)$ (1.1)

Where V is the flow front velocity,
K is the permeability,
n is the resin's viscosity,
dp/dx is the pressure gradient.

For arbitrary reinforcements three orthogonal directions can be taken such that

$$V_X = K_X/n\ (dp/dx),\quad V_Y = K_Y/n\ (dp/dy),$$
$$V_Z = K_Z/n\ (dp/dz) \qquad (1.2)$$

In principle, assuming that the three permeabilities are known, sets of equations can be generated that fulfil considerations of continuity of flow and can be solved by FEA methods. These can predict flow velocity with time in all directions and provide answers to questions of flow front development and fill times.[10, 11, 12] Commercial software of this type is available,[13, 14] so the details of how such models might be developed will not be considered here.

Typical values of permeability can vary dramatically between material types and direction, over about two orders of magnitude between low Vf% random reinforcements (in the region of $1.10^{-9}\ m^2$) to high Vf% unidirectional reinforcements in the transverse direction (in the region of $1.10^{-11}\ m^2$).

It is often the case that the Z direction (through thickness) permeability can be neglected. Even for relatively thick mouldings this is likely to be the case remote from the point of resin introduction, except perhaps very close to the flow front or if mixed reinforcement types are used. Many reinforcements such as balanced woven cloths can also be considered as essentially isotropic in the plane from a permeability viewpoint. Despite their apparent anisotropies, bidirectional cloths generally give essentially circular flow patterns in centre ported tools. Under these conditions simpler models such as those used to model mould filling in thermoplastic injection mouldings may give acceptable information on flow front shape during injection. The use of such models may be adequate for considerations of optimal positioning of in gates, but would be less likely to be useful in predictions of mould filling time. Simple models would also be expected to cope very poorly with point-to-point variations in permeability or easy flow paths.

Simple geometries have been analysed to give analytical expressions for fill time.[15, 16]

From reference 16, assuming in plane permeability is the same in all directions, the general form of the equation can be expressed as:

$$T = CnL^2(1 - \mathrm{Vf})/PK \tag{1.3}$$

where L is a characteristic length of the mould and C is a constant that depends on gating choice. For order of magnitude flow time calculations the following values can be taken.

For an edge gate in a constant width tool C can be taken as 1/4.
For a peripheral gate on a circular panel C can be taken as 1/16.
For a centre ported gate on a circular panel C can be taken as:

$$C = 0.25(X^2 + 2\ln\{1/X\} - 1) \tag{1.4}$$

where X = injection point diameter divided by L. For centre-ported tools rather than narrow rings the X^2 term can be neglected.

For an injection port diameter of 1% of the characteristic length, C is about 2, falling to 0.9 for a port diameter of 10% of L. Thus in rough terms, for a constant pressure resin delivery we would expect peripheral gating to be about four times faster than edge gating and 32 times faster than using a point gate of diameter 1%L. From these simple considerations it would seem that we should always use either edge or peripheral gating. There are, however, good reasons for sometimes choosing to use centre porting, such as: the requirement to seal tools if either of the other approaches are used; the possibilities for easy flow paths with edge injection; and the increased risk of drawing air into the tool through incomplete seals in evacuated tools with peripheral injection.

Another considerations might be how the resin pressure changes with time within the tool. Figure 1.2 shows the pressure distribution within a line gated tool as it develops during the injection cycle.

Also shown is the development of the resin pressure at one point in the tool. Figure 1.3 shows the general form of the pressure distribution development for centre and peripherally ported tools.

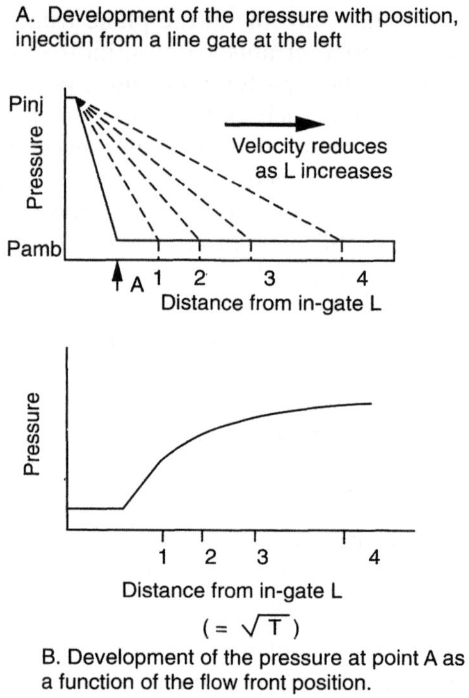

Figure 1.2 For a flat, uniform-width mould with a line gate at one end, injected at constant pressure, the pressure within the mould varies as shown above.

In the most general case:

$$T = n/P\ [f(\mathrm{Vf}, \text{structure})]\ [f(L)] \tag{1.5}$$

K is replaced by some function of the reinforcement's geometry and other properties and L by some function of the tool and gating geometry. The influence of the reinforcement geometry and Vf% on K is considered below.

1.2.3 Influence of reinforcement Vf and structure on permeability

The Kozeny-Carman equation is often taken as the starting point for discussions of how permeability is influenced by the nature of the reinforcement.[17] This is:

$$K = Cd^2B^3/(1 - B)^2 \tag{1.6}$$

Where C is a constant,
d is the fibre diameter
B is the porosity (1 – Vf)

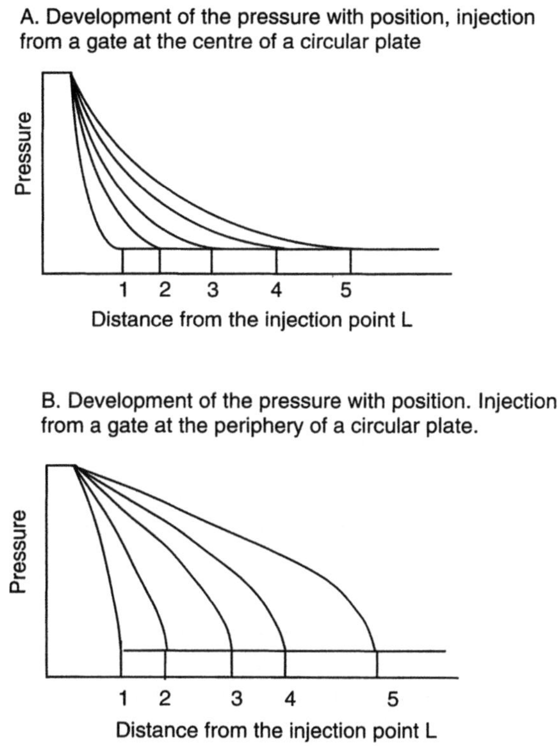

Figure 1.3 The shape of the pressure curve with respect to injection length is a function of the type of gating. Two options are shown here. Peripheral injection leads to the lowest pressure drop and the fastest injection.

As this equation contains a term for fibre diameter and it is known that gross flow is usually dominated by flow between tows rather than between fibres it would be somewhat surprising if the unmodified theory provided a good match to experimental data.

Indeed the match is often poor, as can be seen from Figure 1.4 which shows some measured values of flow rate versus Vf% and the equivalent predictions from the Kozeny-Carman equation. The flow rate data is my own and was obtained for wetting flows using fine woven glass cloth and standard viscosity oil rather than resin. The observed increase in flow time with volume fraction was much higher than that predicted by the Kozeny-Carman model.

An alternative approach is to set the flow rate proportional to the hydraulic radius of the flow channels. At first sight this appears to be more reasonable as it is accepted that the bulk flow primarily occurs between the tows. There is still a problem in that the shape of the flow

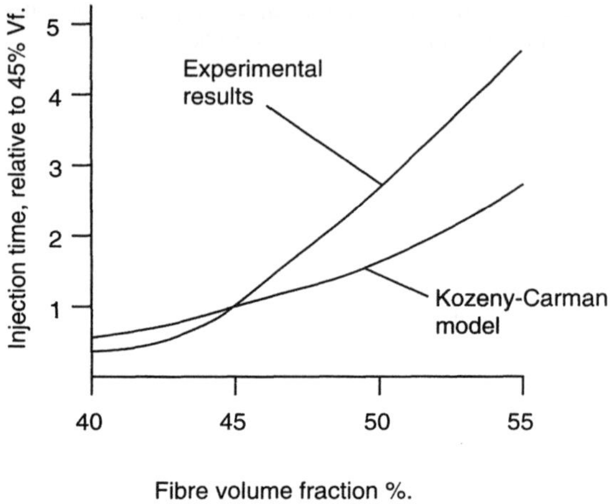

Figure 1.4 Injection time rises steeply with Vf%. Experimental results show a greater dependence on Vf% than the simple theory predicts.

channels changes along their length, so that an element of curve fitting persists. In the case of woven cloth the regions where tows cross will give rise to a different size and shape of flow channel. Even in the case of non-crimped stitched or knitted reinforcement the stitches or loops will tend to lead to differences in flow channel shape and size.

Figure 1.5 shows the flow channels that would be present in the simple case of a 0.90 array of perfectly elliptical tows, perfectly packed together. Even for this simple case the size and shape of the flow channel is very different from point to point. Figure 1.5 also shows the additional complexity of shape introduced if the tow size varies from point to point. The attempt to draw out similar figures for woven cloths taxed my meagre drawing skills to the breaking point.

Figure 1.6 shows a schematic of the flow channels in a plain weave cloth; the figure shows half the thickness of one unit cell. Despite the limitations of my skill the general features should be about right, clearly demonstrating the complexity of the flow channels within even the simplest woven cloth. Of course, in a multilayer stack of cloth the actual flow channel shapes would also depend on the spatial relationships between one layer and the next.

This is shown in Figure 1.7 for a simplified version of the geometry shown in Figure 1.6. It seems likely that without substantial effort in determining the exact shape of the flow channels in differing reinforcements and how these change with Vf% it will not be possible to arrive at an acceptably accurate estimate for permeability from first principles. These considerations lead me to the conclusion that the best that is achiev-

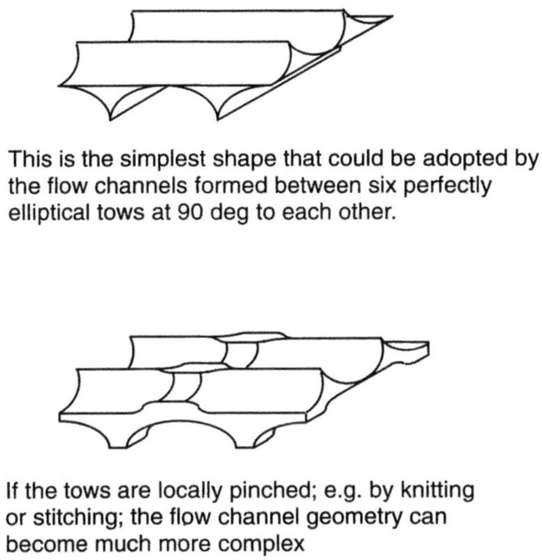

Figure 1.5 The shape of the channels through which resin flows can be quite complex even for relatively simple reinforcements.

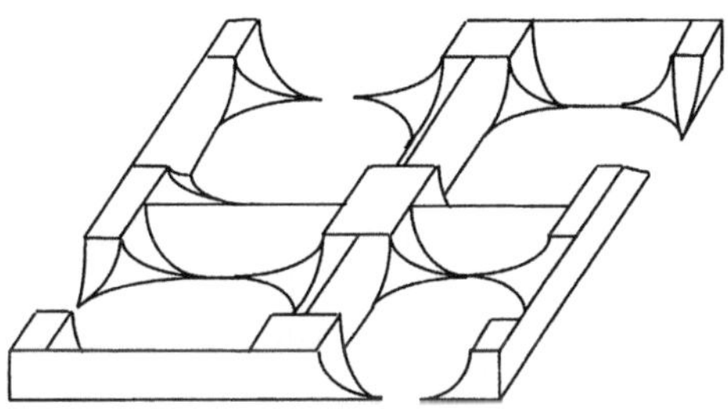

Figure 1.6 Reinforcements such as plain woven cloth can give rise to very complex resin flow channels.

able is likely to be an effective interpolation model that allows estimates to be made for Vf%s between experimentally determined values for a particular combination of reinforcement and matrix. This then raises the question: why not measure the permeability at the required Vf% in the first place? It is also important to measure the permeability with the resin which will be used in production as some experimental studies have indicated differences in apparent permeability for different liquids of similar viscosities.[18] When reinforcements are extensively deformed by mould closure or preforming the point to point permeabilities can be quite

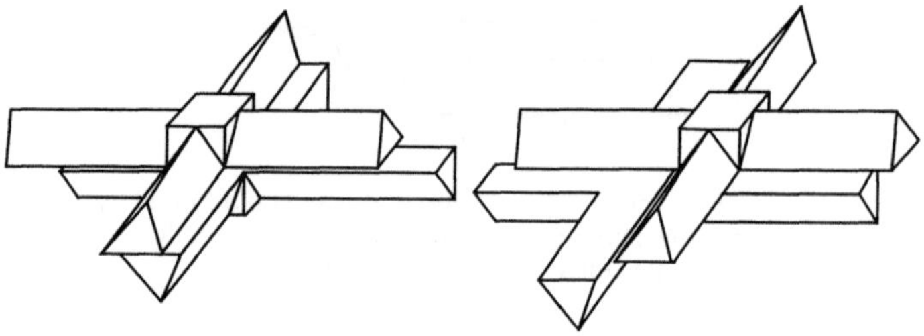

Figure 1.7 The relative positions of flow channels in adjacent plies can also influence the overall flow channel shape (a simplified version of the geometry shown in Figure 1.6 is used here).

different to that of undeformed reinforcement. These changes in permeability may influence resin flow and ideally should be accounted for. In practice, measuring permeabilities for a wide range of deformed reinforcements would be a time-consuming activity and the use of an effective model would be preferred. Some progress has been made in the production of such a model.[19] Having said this, the highly deformed areas of reinforcement within a complex preform assembly tend to be rather small and I have not seen flow problems that could be unambiguously attributed to permeability changes due to deformation of preforms. The other changes in the reinforcement due to preforming deformations, such as thickness or Vf% changes or easy flow path generation, are of great importance and it may be that when these are dealt with the permeability differences become less critical.

All of the above discussion is based on the assumption that the reinforcement geometry gives rise to high and low Vf% regions in some consistent manner. This will be true of the vast majority of reinforcements used in advanced RTM. For materials such as random glass mats of low Vf% it is very questionable whether the concept of a flow channel is meaningful as the cross-sectional area of 'flow channel' is much higher than that of reinforcement tows. The assumptions used above also break down for the more structured reinforcements when they are used at Vf%s below their natural packing fractions. For woven cloth without any pressure applied the thickness of the cloth equates to about 40% Vf. If such a cloth is used in RTM at a lower Vf% the bulk flow is in a sheet between the plies rather than in channels between the tows. If permeabilities are measured for such materials a change in slope can be seen in Vf% versus permeability curves at the point where the reinforcement starts to 'float' in the tool.

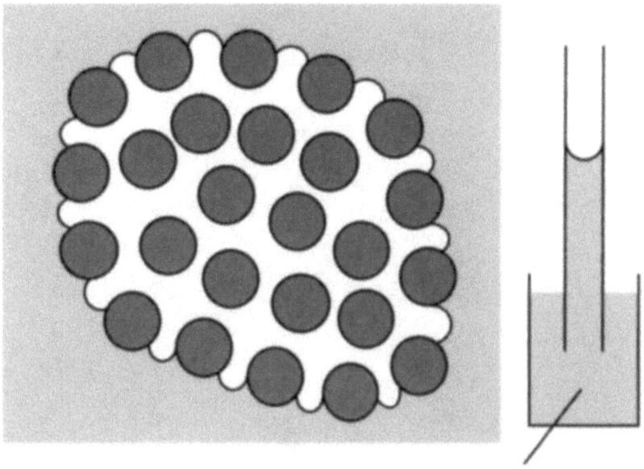

Figure 1.8 When the resin contacts the fibre tow wet-out commences. At the time shown here wetout has just begun and a slight net tension will exist around the tow periphery.

1.2.4 Wetting phenomena

For wet-out to occur naturally the contact angle between an arbitrary surface and a contacting liquid must be less than 90 deg. In the case of epoxy resins and carbon fibres the contact angle would be expected to be below 20 deg and at least partial natural wet-out of the tow is predicted. Ignoring all other forces one can consider what happens when the tow of fibres encounters a wetting liquid, taking a hypothetical case where a volume of resin is radially approaching a tow of fibres. At the point of first contact the contact is tangential between fibres. As the liquid wets the fibres a concave meniscus is formed and the wetting front starts to move inward to minimize surface energy. This point is reached essentially instantaneously and is illustrated in Figure 1.8.

Under the influence of the surface tension the resin continues to move inward until the meniscus is flat and no pressure differential exists across the liquid surface, see Figure 1.9.

A pressure exists across any curved liquid surface, which can be expressed as:

$$P = Y(1/R1 + 1/R2),[20] \qquad (1.7)$$

where Y is surface tension and R1 and R2 are generalized radii. In the case being considered R2 is infinite and R1 can be set at approximately half the fibre to fibre distance. For epoxy resin (surface tension about 40 Dynes/cm) and a fibre to fibre distance of say 0.4 times a carbon fibre radius (approximately equivalent to a Vf of 63%) the maximum pressure

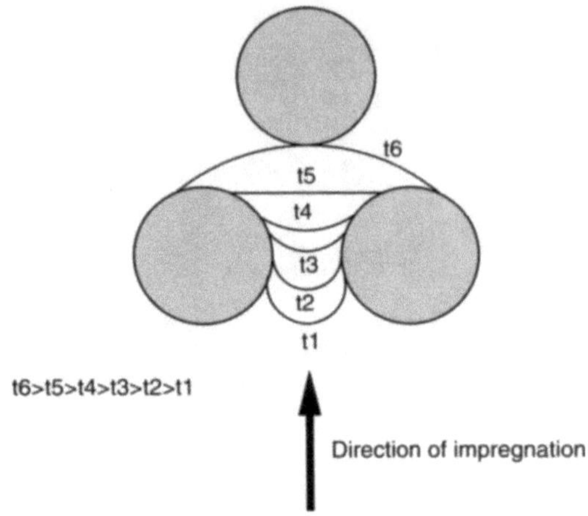

Figure 1.9 Stages in the wetting-out of a bundle of fibres by capillary flow due to surface tension. Wetting proceeds by capillary action until an equilibrium is reached at t5. For further wetting the flow front must be expanded by applied pressure until the next fibre is reached at t6.

across the liquid surface equates to about 28 KPa or 4 psi. Except for positions very close to the in gate, this force greatly exceeds the applied pressure just behind the bulk flow front. In addition to the effect on the flowing resin this force will tend to compact locally the tow by some small but non-zero amount.

If the fibres were perfectly aligned, and neglecting inertia, this pressure would rapidly lead to a close-packed tow and effectively prevent any further resin flow into the bundle (as the transverse permeability of a close-packed array of circular fibres is zero). In real tows the deviations from fibre straightness will give rise to an effective radial tow stiffness to oppose the packing pressure, leaving the tow surface porous and permitting further flow. As before, the flow will cease when an equilibrium position is reached. The only way that natural wetting can continue in the radial direction is if the liquid front touches another fibre prior to reaching the equilibrium position. Whether or not this occurs depends on the contact angle and the fibre/fibre distance. For a uniform array of fibres the dependence of fibre/fibre distance on Vf% can be calculated and used to predict the uniformly packed Vf%s that would give rise to natural wetting at various contact angles (see Figures 1.10 and 1.11).

It is to be expected that in organized reinforcements there will be regions that are naturally wet out by radial flow in the absence of any external pressure. Equally, it is to be expected that there are areas in the

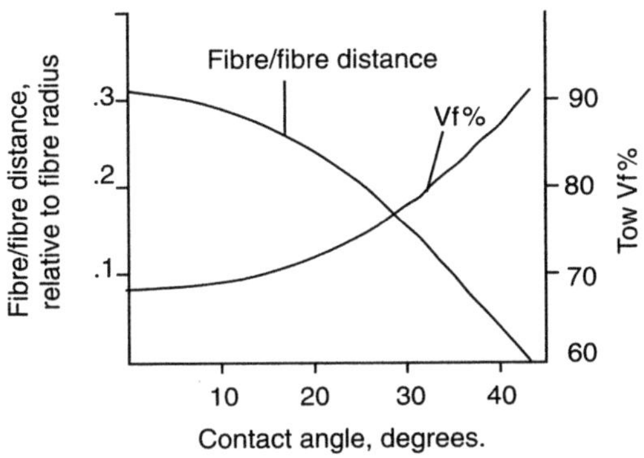

Figure 1.10 For a regular hexagonal array of fibres the maximum distance between fibres that will permit wetting without pressure can be calculated as a function of wetting angle. Distance between fibres can also be converted into equivalent tow Vf%.

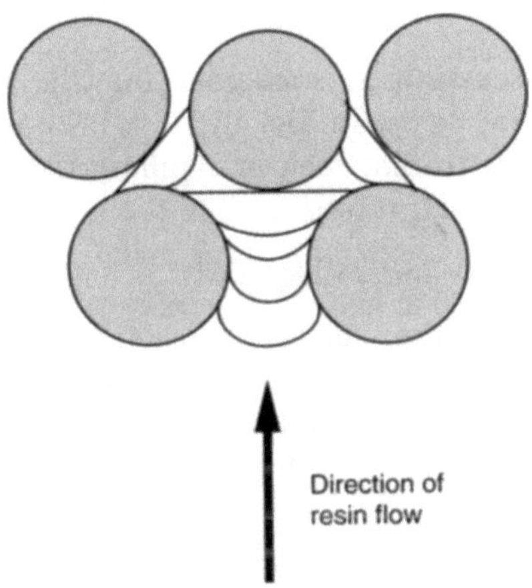

Figure 1.11 Requirements for full wet-out without applied pressure. In regions of high natural fibre packing additional external pressure is not required. The Vf% at which wetting will proceed without pressure depends on the contact angle A (degrees).

For $A = 0$, Vf% = 68%
For $A = 10$, Vf% = 69.2%
For $A = 20$, Vf% = 72.2%
For $A = 30$, Vf% = 78.2%.

Most resins have A between 10 and 20. Tows of fibres will certainly have regions where the Vf% is in the range shown above.

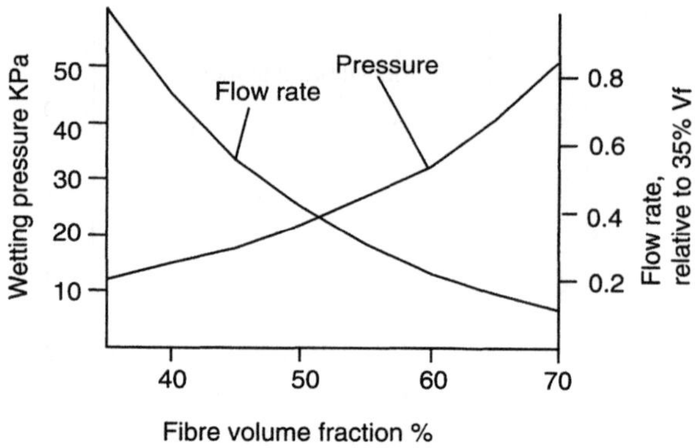

Figure 1.12 Effective wetting pressure in the fibre direction rises rapidly with volume fraction. Permeability, estimated by Kozeny-Carman, falls even more rapidly, so that overall wetting flow rate falls as Vf% increases.

tow that will require additional resin driving pressures to be present to permit full wet-out.

In the above discussions it is assumed that the wetting behaviour gives rise purely to radial flows. In fact there is also an effective longitudinal driving force similar to that seen in capillary rise of fluids in fine tubing. This pressure can be expressed as:

$$P = Y\cos A/(r[1 - \mathrm{Vf}\%]/2\mathrm{Vf}\%),[21] \tag{1.8}$$

For epoxy resins the surface tension will be in the region of 40 Dynes/cm and the contact angle A will be about 18 deg. This leads to a longitudinal pressure of about 14 KPa (2 psi) at 40% Vf and 40 KPa (5.8 psi) at 65% Vf. The pressure at other values of Vf% is shown in Figure 1.12.

These pressures are not high but are real and are in excess of applied pressure in the wetting region. On the other hand the resistance to flow will rise even faster with increase in volume fraction such that wet-out driven by longitudinal flow would be expected to be faster in the lower Vf% regions. Observations on polished microsections suggest that the fibre arrangement within a single tow can be quite variable from point to point, such that the wetting driven flow fronts, both radial and longitudinal, will be rather uneven, presenting opportunities for fine scale air entrapment within the tow (see Figure 1.13).

In addition there will be a tendency for some slight tow compaction in tows that are wet out axially as well as in tows that are wet out radially. Figure 1.14 shows part of such a tow (at about 50% Vf); one effect of the surface tension will be a tendency for the fibres to be drawn together. This can easily be confirmed experimentally by dipping a dry reinforce-

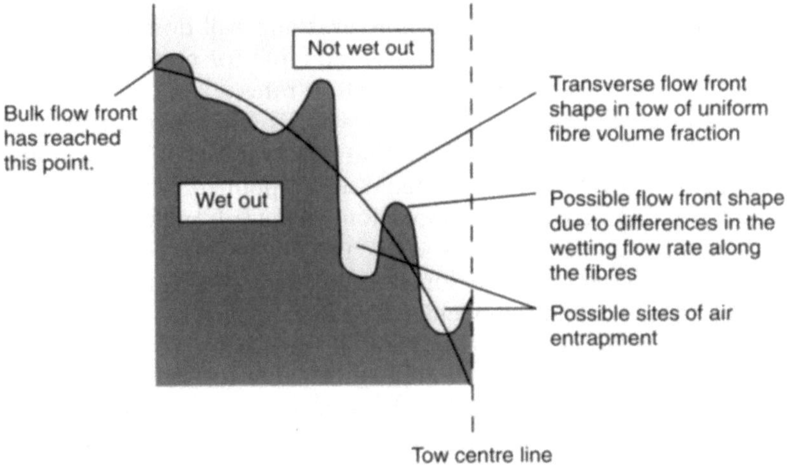

Figure 1.13 Inhomogeneities within the tow can lead to a flow-front shape of complex form and local air entrapment. As drawn the flow is entirely radial; if the general flow front is moving transverse to the flow the probability of air entrapment in an unevacuated tow is much increased as the resin can quickly encapsulate the tow before wet-out is complete.

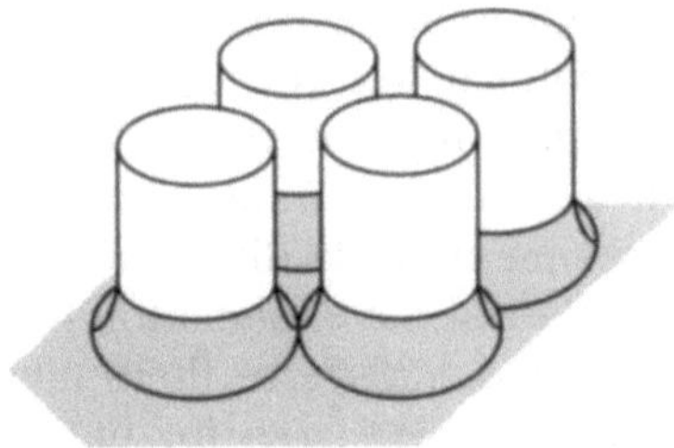

Figure 1.14 The curved meniscus at the fibre surface providc a force tending to draw the fibres together.

ment tow into a wetting liquid. As the liquid is drawn up into the tow the tow can be seen to shrink in diameter. Just as for the radial flow case the effective compaction forces will be small, but not necessarily negligible.

If the wet-out is not very rapid and the point of full wet-out falls behind the bulk flow front the local applied radial pressure will rise to assist in wetting out the tow. In most cases in advanced RTM the wet-out appears to be complete within a few mm of the bulk flow front. In this case it seems most likely that the pressures arising from the wetting behaviour will be in excess of the applied pressures and will dominate the process. This may not always be the case, for example for poorly wetting or very viscous resins, or for very rapid injection rates.

The above notes have considered what is happening at the tow scale and within the tow. It is possible that tow wet-out effects can also have an effect on gross flow behaviour. At the point of wet-out there exists an additional effective tow compaction pressure from surface tension and any local pressure differentials across the tow interface. This pressure will be small, but could lead to some compaction of the tow; which in turn leads to an increase in the open area available for resin flow. For example, if a cloth at 40% Vf consists of tows of fibre at about 50% Vf then 20% of the average cross-sectional area is available for bulk resin flow. If the local pressure increases the tow Vf% to 53% (which seems reasonable from the data shown in Figure 2.5 and the pressures considered above) then the area available for bulk flow increases by about 20%. Little more than 1% increase in mean tow Vf% is required to increase the area available for bulk flow by 10%. Once the flow front has moved on and the fully wet out tows have achieved equilibrium with the local pressure, there will be a tendency for any overcompaction of the tows to relax. It seems quite likely that this relaxation will only be completed after the bulk flow of resin has ceased, so that any local tow compaction pressure effects will be incorporated into the gross permeability measured for the total system.

Reported work on permeability estimation is increasingly looking at the fine structure of reinforcements. It may be that it is not only necessary to look at this fine structure, but also to consider the way that this structure interacts with wetting and transverse flows. It has been noted earlier that in some experiments wide differences are seen in permeabilities for differing liquids. Part of the explanation for such effects may well lie at the scale of wetting and transverse flows and any influences these have on tow compaction and open area for bulk flow. Whether or not wetting flows can influence gross permeability to any great extent they certainly have a strong influence on quality issues such as the entrapment of air on a fine scale within the tow.

1.2.5 Curing phenomena

To a first approximation, the curing process commences at the point at which the various components of the resin are mixed together. (This is not strictly true as polyester resins may use inhibitors to delay the onset of cure until the inhibitor has all been used up and epoxy resins may be unable to cure below some onset temperature.) When the mixed resin is introduced to the tool the first noticeable effect is likely to be a change in viscosity, unless the tool and resin supply are held at the same temperature. Even fairly slight changes in temperature can lead to major changes

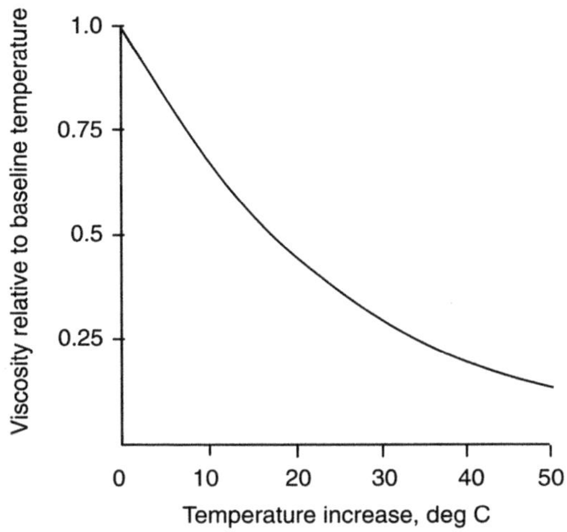

Figure 1.15 Resin viscosity is a very steep function of temperature.

in viscosity as the relationship between viscosity and temperature is exponential in character, i.e.

$$n = n_0 e^{-a(T - To)} \tag{1.9}$$

Where n_o is the viscosity at a temperature of To and a is a constant.

For a value of a = 0.04 °C^{-1}, which is fairly typical for polyester resins,[22] this means that the viscosity will fall by 33% for a 10° rise in temperature and by 56% and 70% for 20° and 30° rises respectively (see Figure 1.15).

The most obvious result of this is that if cold resin is injected into hot tools (or in general terms the temperatures of tool and resin are different) one cannot assume that the use of a fixed resin viscosity in mould flow modelling work will be valid. Ideally, therefore, any mould-filling simulations that are used should account for the way in which the temperature of the resin changes as it is injected. In addition, if curing reactions are initiated during injection the viscosity will change once more. In the long run this change will lead to an increase in viscosity, on a shorter timescale the evolution of exothermic heat as cure proceeds can lead to the resin's temperature increasing, offsetting the viscosity rise. Any viscosity changes from this source should also ideally be included in mould filling simulations.

For most advanced RTM work injection rates are fairly slow and it can be assumed that at any point the resin and the fibre have the same temperature,[23] If the walls of the tool were perfect insulators it would be a straightforward matter to estimate the changes in local resin temperature as injection proceeds, so long as cure exotherms were negligible during injection. Even for composite tools some heat transfer from the tool wall may be important and for heated metal tools heat flows from the tool may have substantial influence, especially on thin components. In the general case it seems most unlikely that assumptions of either purely isothermal or purely adiabatic behaviour will be valid. The estimation of transient heat flows from tool walls and into a cold resin that is spreading through a tool, changing temperature as it encounters fresh supplies of hot reinforcement, will be a very challenging task for the tool designer. Currently, at least one important parameter – the heat transfer coefficient between the tool and the resin – needs to be determined experimentally. See Advani and Bruschke (1994)[24] for more detail. In general, thick components made in GRP tools will tend towards the adiabatic case and thin components made in aluminium tools will tend towards the isothermal case.

The difficulties of acquiring an adequately precise description of the temperature distribution within the tool to permit accurate flow modelling are greatly increased if substantial curing occurs during injection, with the consequent exothermic heat generation and possibilities for both decreases and increases in viscosity. These difficulties are at their greatest for rapidly injected and fast cure polyester resin components, where resin tends to be injected cold into hot tools. In addition to the point-to-point differences in resin temperature across the tool the temperature cannot be considered as constant through the thickness at any point as both conduction from the tool wall and exothermic heat generation will influence the temperature distribution.

One other effect of injecting cold resin into hot tools is that the thermal history of the last resin injected will be quite different from that of the first resin injected. Full cure, and thus demould time, will depend on the cure of the last resin injected; making it difficult to achieve the snap cure at the end of injection that would be preferred in cases of high volume production. Also, differential curing rates across the component could lead to other problems in the cured part such as warpage or surface finish problems. Changing the temperature of the input resin during the injection cycle should enable all these difficulties to be overcome. Having said this, it is not necessarily easy to achieve this in practice. It is possible to inject resin through a separately heated and controlled serpentine gallery so that it picks up heat prior to entering the tool cavity (see Figure 1.16).

It is preferable to use a gallery rather than a heated hose to avoid curing resin in the pipework. The ideal properties of such a resin heating system would be high thermal input capability, rapid control via a feedback loop, low thermal mass and heating the resin in the bulk rather than just at the surface. This is essentially a description of a microwave heating cavity and the use of such a device has been reported.[25]

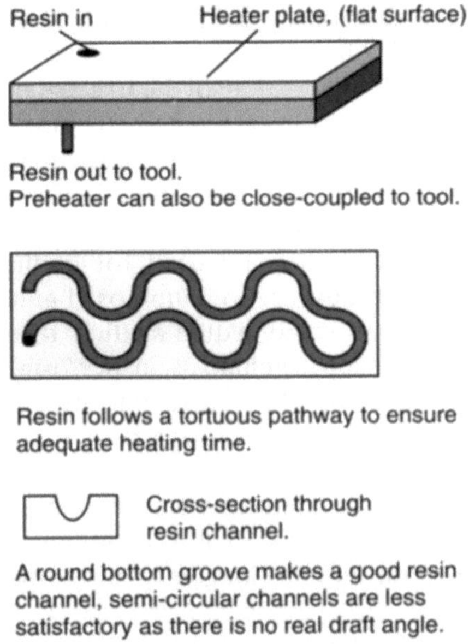

Figure 1.16 To improve control of the process it is sometimes useful to use in-line heaters. Heated hoses can be used, but designs such as that shown below can safely have resin cured within them.

Having said all of the above, the bulk of advanced RTM work is carried out using relatively slowly curing epoxy resins, with resin and tool at similar temperatures, with slower injection rates and high reinforcement Vf%, using fairly thin tool cavities and heated metallic tools. Under these circumstances exothermic heat is limited and resin viscosity can often be regarded as constant during injection for modelling purposes. Even for more rapidly curing systems it may be more realistic to model the flow assuming constant viscosity to identify in and out gate positions. Then a series of instrumented moulding trials can be used to establish the resin's temperature profile, refine the modelling of injection and cure times and set the proper process windows for production engineering considerations.

Once the tool is completely filled the cure of the resin must be accomplished. As before a simplistic overview will be taken rather than any detailed analysis. The details of resin curing are strongly dependent on resin type,[26, 27] only the general features will be considered here. As cross-linking reactions proceed three effects can be seen.

The first is the evolution of exothermic heat as noted earlier. All resins evolve heat in curing. For any given resin the amount of heat to be evolved will be essentially fixed, what can be varied is the rate at which the heat is evolved. Obviously the faster the resin cures the faster the heat is evolved and the higher will be the peak exotherm temperature. In the extreme case the evolution of heat can be sufficient to degrade the resin's performance and even lead to charring. As noted earlier this is most likely in thick, low Vf% components, made in insulating tools. For structural components this has some important implications. If a component must be subjected to qualification testing the cure rate should be considered as fixed at the rate used for qualification mouldings. If substantially faster cure rates are to be imposed at a later date the mouldings should be subject to some requalification to ensure that increasing peak exotherm has not led to changes in performance. The exothermic heat will normally peak on the centre line of the component so, as a minimum, the interlaminar shear strength of the component should be checked at the new cure rate. Even for more commercial mouldings the cure rate should not be treated as a free variable that has no influence on moulding quality. The peak exotherm temperature can also have a profound influence on mould longevity if temperature sensitive tool materials are used.

The second effect is the increase in viscosity. As the resin cures the viscosity will increase as the resin's molecular weight increases. As the viscosity increases a point is reached where the resin gels and can no longer be considered as a liquid. The exact definition of the gel point seems to be a subject of some discussion,[28] but from a functional viewpoint it can be considered as the point at which flow becomes impossible. Beyond this point cross-linking continues in the solid state but at a slower rate than previously experienced. The gelled material can be considered as having a rubbery nature. If the partially cured gel were to be cooled it would lose this nature becoming a glassy solid at some temperature defined as the Glass Transition Temperature or Tg. As cure proceeds in the solid state the Tg will be increasing. Depending on the particular system the Tg may eventually exceed the cure temperature or may level out below the cure temperature. Polyester and vinyl ester resins tend to have Tgs at or above cure temperature, as do phenolics and tetraglycidyl epoxy resins. Diglycidyl epoxy resins tend to show Tgs below the cure temperature. The reasons for this difference seem to lie in the fact that the first group of resins have a greater number of reactive sites and thus a higher probability of continuing reaction in the gel and glassy states.

The last effect is that the resin will be shrinking. This shrinkage will start in the liquid state and continue after the resin has gelled, until the chemical reactions cease. Data on shrinkage are sometimes available from resin manufacturers, however only the total shrinkage is likely to be quoted and the proportion of the shrinkage that takes place in the liquid and solid states is very seldom available. The simplest way of estimating resin shrinkage would be to measure the density of cured and uncured

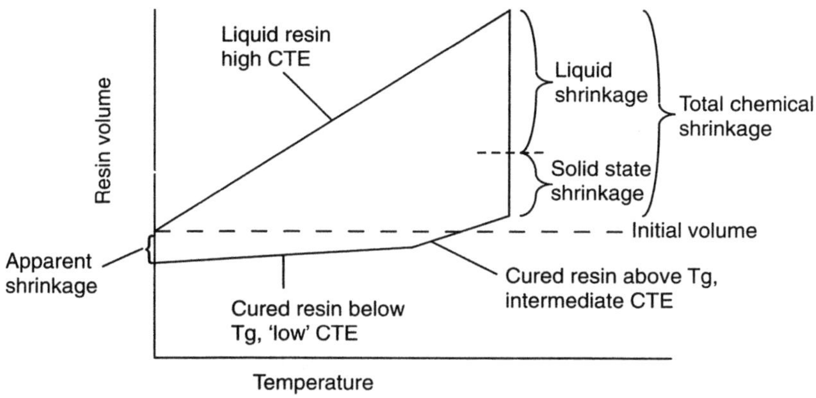

Figure 1.17 The low apparent shrinkage of hot cured epoxy resins may disguise a much higher true total chemical shrinkage.

materials and ascribe this difference to shrinkage; assuming no weight loss on curing. This might be termed the apparent shrinkage. Standard tests for resin shrinkage tend to use a casting mould of fixed dimensions, the shrinkage being calculated between the cured resin dimensions and the mould dimensions at room temperature.[29] What such a test will indicate is the total of solid state shrinkage from the point at which the curing resin becomes self-supporting added to the thermal shrinkage from the cure temperature down to room temperature. This does not tell the whole story, especially for thermally cured resins. As the resin is heated it expands at a thermal expansion rate far higher than would be seen in the cured resin, either above or below Tg. On cure the resin will begin to shrink in both liquid and solid states giving a total resin shrinkage as shown in Figure 1.17.

From this it can be seen that neither the apparent shrinkage nor the shrinkage measured from a casting give a true measure of the total chemical shrinkage. A combination of rheological measurements and the use of techniques such as mercury bulb dilatometry would be able to give a better picture of total chemical and thermal shrinkage and its development during cure in the liquid and solid states. The liquid phase shrinkage can be of importance in the cure of RTM components in rigid tools, sometimes leading to voidage if the shrinkage cannot be fed by supplies of additional resin. The solid state shrinkage can lead to poor surface finish and perhaps to defect formation in areas of totally constrained resin (by virtue of the resin's high bulk modulus even though its tensile modulus is very low at this point in the cure). To conclude, resin shrinkage can have some important effects in advanced RTM work; but the data required to explicate these effects are usually not available. Some polyester resins are sold with low profile additives which expand during cure to offset the cure shrinkage. These seem to be effective in terms of surface finish improvement but require elevated temperature cure such that some of their advantages may be lost as thermally induced cure stresses may be increased.

As noted earlier, the understanding of RTM flow and cure phenomena is of great assistance in tracking down and correcting the causes of quality problems in RTM production. In addition to the understanding of flow there are other, more technological, factors that can influence quality. For this reason the consideration of how to approach defect reduction exercises will be postponed until some of these issues have been covered. Chapter 10 discusses some of the many potential defect causing mechanisms that can be experienced in the development and manufacture of RTM components.

REFERENCES

1. Molnar, J. A., Trevino, L. and Lee, L. J. (1989) Mold filling in structural RIM and resin transfer molding, *44th Annual Conf. Composites Institute*, SPI, Feb. 1987, Session 20A.
2. Lundstrom, T. S. (1994) Void formation and transport in the RTM process, *Proc. 39th International SAMPE Symposium*, April.
3. Lekebou, C., Johari, M., Norman, D. and Bader, M. (1996) Measurement techniques and effects on in-plane permeability of woven cloths in RTM, *Composites*, part A, 27.5.1996, 401–8.
4. Hannoura, A. A. and Barenda, F. B. J. (1981) Non-Darcy flow: a state of the art, *Proc. Euromech.*, 143, Delft, The Netherlands.
5. Injectex RTM: a solution to increasing cost-effectiveness of composites, *Composites News* **3**, 1994, Ciba Geigy, 8–9.
6. Chick, J. P. and Rudd, C. D. (1996) Material characterization for flow modelling in structural reaction injection moulding. *Polymer Composites* **17**(1), 124–35.
7. Chan, A. W., Larvine, D. E. and Morgan, R. J. (1993) Anisotropic permeability of fibre preforms. Constant flow rate measurement, *J. Compos. Mater.*, 27.996.
8. Yates, B., McCalla, B., Phillips, L., Kingston-Lee, D. and Rogers, K. (1979) The thermal expansion of carbon fibre reinforced plastics. Part 5. The influence of matrix curing characteristics, *J. Mat Sci.* **14**, 1207–17.
9. Darcy, H. (1856) *Les Fontaines publiques de la ville de Dijon*, Paris: Delmont.
10. Chan, A. W. and Morgan, R. J. (1994) Computer modelling of liquid composite molding for 3-dimensional complex shaped structures. In *Proc 10th ASM/ESD Advanced Composites Conference/Exposition*, Dearborn, MI: ACCE, Nov., 341–5.

11. Lee, L. J., Young, W. B. and Lin, R. J. (1994) Mold filling and cure modelling of RTM and SRIM processes, *Compos. Struct.* **27**, 109.
12. Gauvin, R., Trochu, F., Boudreault, J. and Carreau, P. (1993) Finite element simulation of the resin transfer molding process, *ECCM 6*, Sept., Woodhead Publishing Ltd, 57–62.
13. Rudd, C. D. and Long, A. C. (1994) Design strategies for fibre preforms, *Proc. 10th Annual ASM/ESD Advanced Composites Conference/Exposition*, Dearborn, MI: ACCE, Nov.
14. Tucker, C. L. and Virlouvet, P. H. (1991) *TIMS Users Manual Version 2.1*, Polymer Processing Laboratory, Dept of Mechanical and Industrial Enginering, University of Illinois, Urbana.
15. Cai, Z. (1992) Analysis of mold filling in RTM process, *Journal of Composite Materials* **26**(9), 1310–38.
16. Gebart, B. R., Gudmunson, P. and Lundemo, C. Y. (1992) An evaluation of alternative injection strategies in RTM, *Proc. 47th Annual Conf. Composites Institute*, SPI, Feb.
17. Carman, P. C. (1937) Fluid flow through granular beds, *Trans. Int. Chem. Eng.* **15**, 150–66.
18. Williams, J. G., Morris, C. E. M. and Ennis, B. C. (1974) Liquid flow through aligned fibre beds, *Polymer Engineering and Science* **14**(6), June, 413–19.
19. Rudd, C. D., Long, A. C., McGeehin, P. and Smith. P. (1996) In plane permeability determination for simulation of liquid composite molding of complex shapes, *Polymer Composites* **17**(1), Feb., 52–9.
20. Moore, W. J. (1962) *Physical Chemistry*, London: Longman.
21. See reference 18 (Williams *et al.*, 1974).
22. Scott, F. N. (1988) Processing characteristics of polyester resin for the resin transfer moulding process, PhD Thesis, University of Nottingham.
23. Lin, R., Lee, L. J. and Liou, M. J. (1991) Non isothermal mould filling in RTM and SRIM, *Proc. 49th Annual Technical Conference*, ANTEC 91, SPE.
24. Advani, S. G. and Bruschke, M. V. (1994) Resin transfer moulding phenomena in polymeric composites. Chapter 12 of S. G. Advani (ed.), *Flow and Rheology in Polymer Composites Manufacturing*, Amsterdam: Elsevier Science BV.
25. Johnson, M. S., Rudd, C. D. and Hill, D. J. (1995) Cycle time reductions in resin transfer moulding using microwave preheating, *Proc. Inst. Mech. Eng.* **209**, 443–53.
26. Kenny, J. M., Maffezzoli, A. and Nicolais, L. (1990) A model for the thermal and chemorheological behaviour of thermoset processing (II) unsaturated polyester based composites, *Composites Science and Technology* **38**, 339–58.
27. Loos, A. C. and Springer, G. S. (1983) Curing of epoxy matrix composites, *Journal of Composite Materials* **17**, 135–69.
28. May, C. A. (1988) *Epoxy Resins, Chemistry and Technology*, New York: Marcel Dekker, Chapter 14.
29. ASTM D2566. Linear shrinkage of cured thermosetting casting resins during cure.

2 Materials for RTM

2.1 REINFORCEMENTS

2.1.1 Fibre types

Fibre types known to have been processed by RTM include E, R & S glass, quartz, a wide variety of high-strength and high-modulus carbon fibres and aramids. No difficulties are known with any specific fibre type. For maximum quality in RTM moulding near instantaneous wet-out by resin is required. The resin needs to wet the reinforcing fibres naturally. Some difficulties have been experienced with glass fibre and cold cure phenolic resin combinations, but most resin-fibre combinations do not present difficulties. A simple check to verify that resin is naturally wicked up by the reinforcement at the injection temperature to be used is a sensible precaution to ensure compatibility at this level. This need be no more complex than placing a small quantity of resin on the reinforcement surface at the required temperature; if the resin is rapidly drawn into the reinforcement compatibility should be adequate.

2.1.2 Reinforcement types

As noted earlier a wide variety of reinforcement types are readily processable, some of these are noted below.

(a) Unidirectional

In some cases UD tows have been used as local reinforcement, either bare or contained in a braid or other carrier to improve handleability. Preforming techniques that rely on tow laying to build up the reinforcement pack have also been investigated.[1] It is more common to use broadgoods in a variety of forms.

Woven UD materials are widely available and easily processed. Gaps between tows can act to increase injection rates. Unless UD woven materials are bound in some way they have a tendency to fray badly which makes control of the preform rather difficult. Powder binders have been used to minimize this problem as have the use of a tacky warp thread and a separately applied tacky web applied to one surface of the cloth.

Warp and weft insert knitted UD forms are available in which the UD tows pass either along or across the cloth. Some knitted UD forms are also prone to fraying when trimmed; others use a lock stitch that reduces the problem. The tow shape is to some extent controllable, some UD knits have essentially flat tows while in others the tow cross section is nearly circular. This flexibility permits some measure of designing the reinforcement for maximum permeability.

Other methods for increasing permeability that have been seen in UD woven and knitted cloths include lightly overwinding some tows, using tows of different tex values or omitting occasional tows altogether. All these methods aim to increase the available volume for resin flow and can also be applied to bidirectional forms (see Figure 2.1).

Commercial materials of increased permeability are available.[2] The improvements in permeability appear to be accompanied by a reduction in strength properties, especially compressive strength.[3] To offset this one might suggest that the flow-enhanced cloth should not be used for every ply. The use of such reinforcements would require a careful balance to be taken between any reductions in processing costs, increased inventory costs, the possibilities for layup error when used on a partial replacement basis and any reductions in mechanical properties.

Unidirectional broadgoods based on the use of a light web of thermoplastic fibrils or regular strips of a heavier non-woven material have also been commercialized.[4] The fibril bound material is very uniform and resistant to fraying, but would be expected to have a lower permeability than the woven or knitted forms. The material that uses spaced strips frays very badly if cut anywhere other than at the strips.

(b) Bidirectional forms

Conventionally woven cloth is the most commonly used reinforcement type, and is available in a staggering variety of materials, weaves and weights. The range of cloths usable in RTM cannot be specified exactly, but cloths from 1 K carbon to glass cloths over 1000 gsm have certainly been processed. In aerospace work 3 K and 6 K carbon cloths are generally specified. Most aerospace cloth specifications cover plain, crowsfoot and satin weaves. Unbound cloths are easily distorted, with plain weaves generally the most stable and satins the least. Any of these weave types

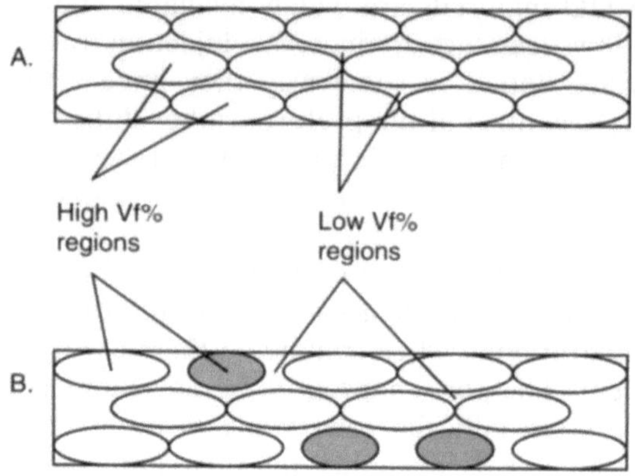

Figure 2.1 If tows are constricted by twisting or overwrapping, then the area of low Vf% regions that act as resin pathways is increased. A shows standard weave. B shows one layer with one tow out of five treated, one layer of standard weave and one layer with two tows out of five treated.

can be stabilized by using a powder or fibril web binder. If mechanical preforming is to be used some sort of binder is mandatory and the formability of the cloth may be an important factor. Cloths can be woven and are commercially available that have exceptional deformation characteristics, these can be of value in reducing the number of pieces of cloth required to cover any given shape. The question of cloth deformability and the associated changes in warp/weft angles and ply thickness can be very important in RTM, especially with rigid tools. These issues are covered in more detail in section 3.2.

A variety of non-crimped cloth constructions are available that aim to reduce or remove the crimping of fibre tows inherent in woven cloths and thus improve mechanical properties, these can also be used in RTM. Fine threads can be woven together with the warp and weft threads held between them. These are much more prone to distortion and fraying when cut than conventionally woven types. Warp/weft insert knitted cloths are available and similar comments apply to these as to the UD forms mentioned earlier.[5] In addition care has to be taken that the knit pattern used does not interfere with the cloth's ability to form on a 3D surface without wrinkling of the tows. The understanding of the deformation behaviour of knitted cloths is currently inferior to that of woven cloths and more trial and error may be required in defining developed shapes etc. (the developed shape of a 3D formed preform is that shape which an accurately trimmed preform would take up if returned to the flat state).

(b) Special purpose reinforcements

The flexibility of RTM to handle a wide variety of reinforcement types is one of its great strengths, but care has to be taken with the selection of the more unusual reinforcements to ensure the correct balance of materials and labour costs and performance.

Multidirectional materials

Stitching, weaving and knitting can all be used to produce broadgoods with three or more tow directions in the plane of the reinforcement, knitted types are now probably the most common.[6] Arrangements such as 0,90,+/–45 are available and can reduce layup costs in some cases, although another cloth with 0,90,+/–45 may be needed for a balanced and symmetrical laminate. These multiaxial knitted cloths deform by slip of the fibre tows through the knit loops and in this way can achieve surprisingly high levels of deformability. This deformation mode is not well understood and the use of such materials to form 3D geometries would have to be on a trial and error basis.

Shape weaving

Jacquard looms are available in which each warp yarn is independently controllable and can be raised or lowered to insert the weft yarn. Such looms are routinely used to produce patterned fabrics and can also be used to produce reinforcing fabrics. Most commonly these are used to produce what is in effect a two-layer fabric where the two layers are brought together at the edges of the required shape. Radomes can be woven such that the shape appears as a triangle when flat on the loom but can be opened out to give a cone (see Figure 2.2).

The excess warp threads must be cut away before the material can be used. Only shapes that can be developed from flat to 3D shapes are possible. Traditionally, jacquard looms are controlled by punched cards in the same way as early computers or player pianos. Electronic versions are now available which ought to reduce set up costs. Jacquard looms put more stress on the fibres being woven and thus tend to run rather slowly compared to broadgoods looms, adding to production costs.

Shape knitting is much more flexible in terms of geometry but is limited to plain knitting (i.e. without inserted UD tows). This means that only low Vf%s are possible. The small loops formed in plain knitting are very damaging to stiff, low-extensibility fibres which restricts the process to glass and aramids, although spun-staple carbon fibre can be knitted. The oxidized PAN precursor to carbon fibre behaves as a textile fibre and can be knitted. If such knitted preforms are heat-treated then carbon fibre will result, although the mechanical properties would be rather limited and perhaps best suited to carbon-carbon component manufacture.

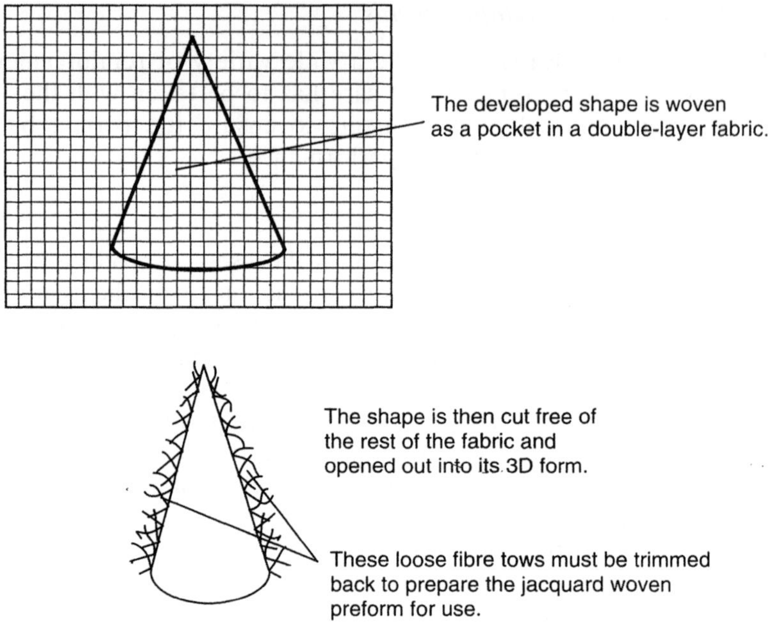

Figure 2.2 3D shapes can be created from flat woven material using jacquard looms.

Spiral weaving

It is possible to weave on a curved warp path such that, as weaving proceeds a spiral of cloth emerges with radial wefts. These reinforcements could be used to make objects of circular symmetry such as discs, or cones for various applications (e.g. flywheels, road wheels etc.). In the simplest process the wefts will be more widely spaced on the outside of the spiral than on the inside. Additional wefts can be inserted on the outside part of the spiral to limit this effect. The set-up and weaving costs are much higher than for plain weaving and a balance must be sought between layup cost reductions and higher material costs.

Continuous sections

In a jacquard loom each warp tow is individually controlled. In the weaving of continuous sections control is exercised over groups of tows, greatly simplifying the process. In general, any shape that can be folded flat without distortion or stress in the fibres can be woven as a continuous section.

Figure 2.3 shows a variety of possible types, from simple 'T' or 'H' sections to woven sandwich structures.[7] These can be used within an RTM preform for such tasks as ensuring fibre continuity in areas of complex geometry etc.

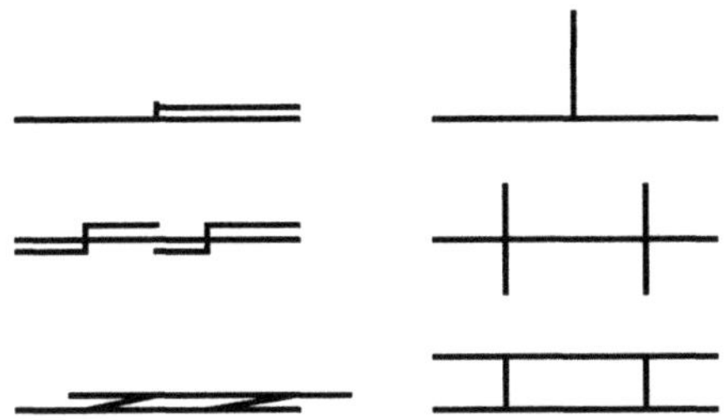

Figure 2.3 A great many shapes can be woven as continuous sections, the basic requirement is that they can be folded into a flat form.

Three-dimensionally reinforced materials
In general terms the function of 3D materials is to suppress delamination in the through thickness direction by introducing fibres in this direction. They range from thick, multilayer cloths produced on near standard looms, to orthogonally arranged blocks with three fibre directions, to shape woven or braided articles with a multiplicity of fibre directions. In the simplest case multiple warp layers are used and some of the weft fibres pass through and interlock several layers of warps providing an element of through thickness reinforcement. Other weft fibres are held straight so that perhaps 80% of the fibres are straight and 20% are used to interconnect the layers. These materials are claimed to have high impact properties and should be able to give rapid layup as one ply might replace many plies of conventional cloth. On the other hand, in-plane properties would be expected to be reduced and it is relatively uncommon to have large areas of constant thickness in real components.

The true 3D materials offer the highest levels of delamination resistance and hold out the possibility of net-shape preforms that require no layup labour.[8, 9] Much of the development work has been carried out for carbon\carbon composites of very high value. It is unclear whether the current low preform production rates can be increased to the point that these would find major applications in lower added value markets. The main relevance of such materials to RTM lies in the fact that they are very hard to process by other moulding techniques and RTM is often the preferred process.

An element of 3D properties can also be achieved by stitching with aramid fibres,[10] or by the use of 'stitched effect' fabrics in which aramid fibres pass through several layers of carbon fibre.

For all these materials the multiplicity of fibre directions tends to reduce the overall Vf% somewhat and the in-plane properties would be expected to be below those for woven cloths. In addition, for the orthogonally woven materials (or others with more fibre directions) the composite as

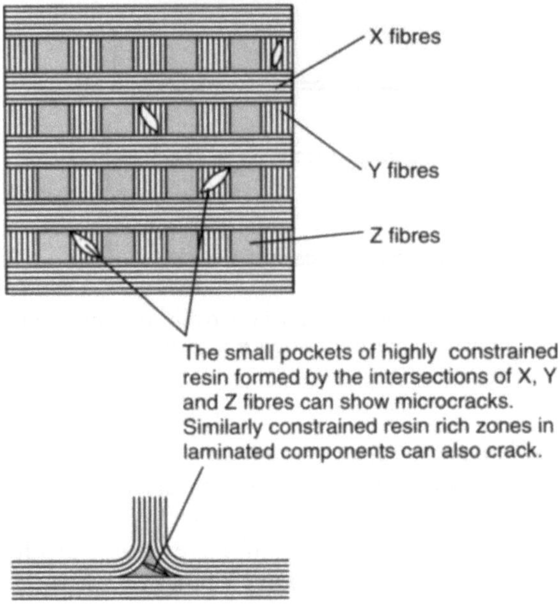

Figure 2.4 Resin pockets that are constrained by the fibres can crack as a result of thermal stresses.

a whole is constrained against through-thickness shrinkage by the fibres. For laminated materials much of the resin's curing and thermal shrinkage stresses are relieved by through thickness shrinkage. If this is prevented then any internal resin rich zones can crack as a result of these stresses; such resin rich zones are commonly caused by the spatial interaction of multiple fibre directions.

Figure 2.4 shows the appearance of such cracks as observed by the author in a thick orthogonally woven carbon fibre block made by RTM. It is worthy of note that the resin used in this case had proven itself to be very resistant to microcracking in conventional laminates under extreme temperature cycling, i.e., no microcracks after 50 cycles between 100 °C and liquid nitrogen.

Pile fabrics

It is possible to raise a pile on the surface of a fabric by a variety of techniques from brushing to weaving on a velvet loom. Such fabrics may be termed 2.5 D and are claimed to give improved interlaminar properties,[11] although whether this is a function of the pile fibres themselves or a thicker interply resin layer (as shown in Wang, 1989[12]) is unclear. Long-pile fabrics might find applications in fire resistance as the pile would be expected to stabilize the charred surface.

Felt materials
These are random in the plane and low Vf%, and contribute little structurally. They may be used as surfacing tissues, including metallized varieties to give the required electrical properties.[13] Thicker felts have a very high bulk factor (ratio of compacted and uncompacted thickness) and are sometimes used as spring-like packers to ensure that the structural reinforcement is held against the toolface in complex layups. For example, on a tapered flange a thick felt may be used to avoid the labour associated with tapering off many plies of reinforcement. It should be noted that these felts are often made with hydroscopic binders such as PVA; these must be thoroughly dried before injection to prevent processing problems.

It is to be expected that the range of types of reinforcements will continue to increase and that for many of them RTM may be the only practical processing technique. For some of them it would be expected to be very difficult to generate allowable properties and to my knowledge none of the special purpose reinforcements is generally qualified for aircraft use. This is not, of itself, a problem but the costs of qualification for a new material form can be high and must be accounted for in any component development costing.

2.1.3 Relationships between volume fraction and applied pressure

These relationships have important consequences for rigid tool, fixed cavity, RTM in terms of component and tool design and it is important that they are taken fully into account.

For any reinforcement, as applied pressure increases Vf% increases, in a way that is roughly proportional to the log of applied pressure. There is a natural packing density at zero applied pressure which can range from about 10%, or even less, for felts to around 40% for bidirectional woven cloths to 50% or more for UD materials. With additional pressure the Vf% rises most rapidly for the random material and least rapidly for the UD material.

Figure 2.5 shows fairly typical figures for UD, woven and random materials. In addition, if the pressure is applied and then released (for example in a preforming operation) there may be substantial hysteresis and the new zero pressure Vf% can be substantially higher than that prior to any compaction. Any decay of this increased level of zero pressure Vf% with time after the removal of the consolidating pressure would have to be assessed on a case-by-case basis. This effect could be of importance in some cases. For example, if a part were made with pressed preforms and the preforms were loaded immediately into a tool it is possible that low-resistance pathways could exist within the tool if the new zero pressure Vf% leads to a reinforcement thickness below that

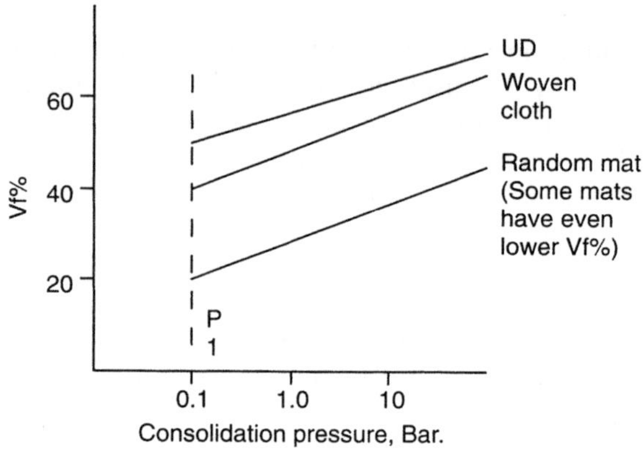

Figure 2.5 Consolidation pressure is a very strong function of Vf% and has some important consequences for RTM. At pressures below P1 the reinforcement is barely constrained by the tool. Movement of the reinforcement due to resin flow is likely and the resin will tend to flow between the plies in an uncontrolled manner. The maximum consolidation pressure depends on the type of tooling used. Vacuum bag tooling is limited to 1 bar. Even for rigid tooling the required loads rise so quickly that pressures very much above 1 bar are seldom used. 1 bar = 10 tonnes of load per square metre of tool face.

of the tool cavity. In this case the injection time, and quality, might be strongly influenced by the preform storage time prior to use. These factors and any influence on them of binder use, or storage and use temperatures are not well understood, but could lead to an unexpected source of process variability.

Other effects may be of more immediate concern. For example, for a woven cloth as shown in Figure 2.5 to achieve 55% would require a pressure of around 0.1 MPa. If four plies are required but five are used the Vf% would rise to nearly 70% which would require a closing force almost 100 times higher than that expected. Only 1% of the total tool area need be affected before the mould closure forces are doubled. It is easy to see how this could happen if different areas of the tool require different ply counts and the position of the extra plies is poorly controlled. Fixed-cavity RTM in rigid tools requires close control of fibre packing and is very intolerant of variations, especially if the target Vf% is close to the maximum and/or the moulding is thin. To prevent such problems from occurring a somewhat lower Vf% target can be used, although not so low that the reinforcement 'floats' in the tool cavity.

To a first approximation, problems of reinforcement movement under injection conditions may occur if the injection pressure is much higher than the applied packing pressure. This may act to set the minimum usable volume fraction of the reinforcement. Another way to control problems

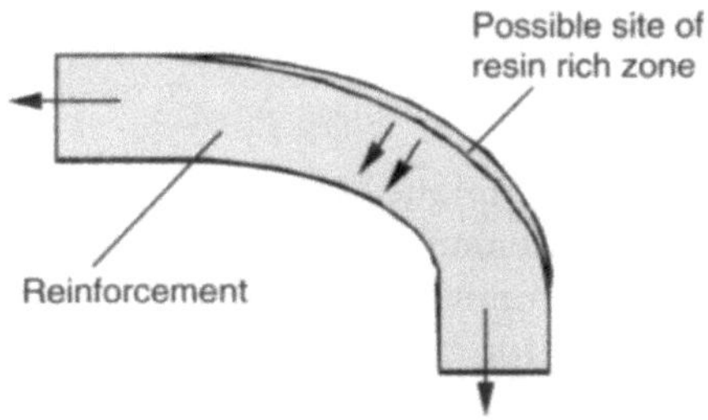

Figure 2.6 When rigid tools are used resin rich zones will be formed if the reinforcement does not fill the mould cavity exactly. If reinforcements are formed by mould closure then the mould closure forces can be reacted by the reinforcement, shown by the two large arrows. At the corner the resultant of these forces, shown by the two small arrows, acts as an additional compaction force pulling the outermost reinforcement ply away from the tool wall and generating a gap. This gap will produce a resin rich zone and may also act to distort the flow front grossly, leading to air entrapment in unevacuated tools.

of excess local Vf% in fixed cavities is to take ply drop positional tolerances wholly on the negative side such that even at top tolerance excess fibre will not be present. In this case, if the ply drops are at the surface, resin-rich layers can be produced which can be a problem in themselves. The use of felt packers or foam cores can also assist in avoiding these problems. In addition, when a reinforcement is preformed or deformed to fit a tool surface its thickness will change, which can lead to similar problems. This is more thoroughly dealt with in Chapter 3.

Lastly, when a reinforcement is formed around a radius by the action of mould closure a tension can be generated in the reinforcement. This equates to an additional packing pressure and can lead to the reinforcement's not being in contact with the outer toolface (see Figure 2.6).

Resin rich zones can then be created and these can additionally act as easy resin flow paths distorting the desired shape of the flow front. Well-made preforms seem to be less subject to this effect than reinforcements formed by the mould closure and the use of preforms can help to overcome these problems. Essentially, the tool needs to replicate how the reinforcement will behave, rather than expecting the reinforcement to follow the shape of the tool automatically. It may be necessary to modify tool geometry to follow the actual shapes made by the formed reinforcement.

All these effects can be lessened by using non-rigid tooling, at the cost of losing some control over the thickness. For complex geometries, rigid tooling may be a necessity, especially if multipart tools are required. On a more positive note, the force needed to close a rigid RTM tool can be used as a partial check that the layup is correct, either very low or very high closing forces can act as an indicator that something is amiss.

2.2 RESINS FOR RTM

2.2.1 Introduction

Whatever the processing technique the matrix resins must match both the process and use requirements. The use requirements in terms of mechanical properties, environmental resistance, costs, storage, safety etc. are more or less the same for all processes and will not be considered here. Some resins that have toxicity problems when handled may be better processed by RTM as contact between personnel and uncured resin can be minimized; but only by the use of properly designed equipment and operating practices. This last point must be emphasized as I have seen many examples of RTM production areas that were very badly contaminated by the resins used. The mere selection of some form of RTM operation is no guarantee of a lack of contact between resins and operators. In RTM we take low-viscosity resins which may be of high reactivity and/or at an elevated temperature; we then apply pressure to the resin. Without a lot of care there are many opportunities for spreading the resin far and wide around the plant. This usually occurs as a fairly slow drip of resin from leaking seals, but I have seen resin spraying from leaks in pipework or blowing back out of poorly designed tools and, even worse, flushing solvents being splashed around to create a mist of acetone carrying reactive chemicals.

In principle, so long as the resin's cure can be delayed until the mould has been filled, without excessive driving pressure being used, that resin is usable for RTM; at least with that mould/preform. In practice a viscosity below about 10 poise is preferred and many RTM resins have much lower viscosities. It is not absolutely necessary to delay the onset of cure until the mould is full, and many high-speed examples of RTM use resins that are curing as they are injected. Modelling flow, heat transfer and cure in these systems can be quite difficult. When aerospace epoxies are used it is more common to inject the resin at a temperature where little reaction occurs during the injection phase, the mould being raised to the cure temperature after injection.

Figure 2.7 shows typical viscosity/time isothermals for an epoxy and demonstrates the variety of process windows possible, and thus by implication the range of viscosities.

The formulation given above merely indicates whether or not a given resin could fill a tool in the necessary time. To ensure adequate quality two other considerations need to be added. The first is that the resin must wet the reinforcement as quickly as possible to prevent some modes of void formation. The second is that volatile (or more correctly gaseous) species should not be evolved by the resin after mould filling or during cure. Any gases would be unable to escape from the mould and voidage

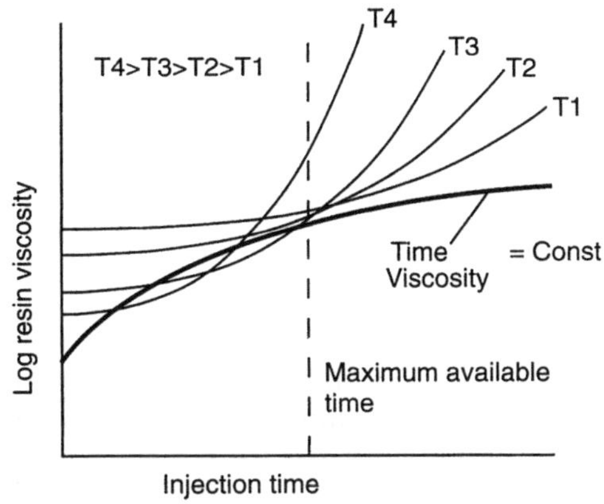

Figure 2.7 Selection of process conditions from resin viscosity data. Maximum available time is chosen from process engineering considerations. Injection time is shortest at T4. There may just be a process window available at T2. No process window is available at T1. This assumes that the shape of the time versus viscosity curve is known for the tool and reinforcement. This is often not the case and trial and error methods are used.

would result. It should be noted that resins that contain a high proportion of volatiles, such as polyester resins, are perfectly usable in RTM unless very high levels of in-mould vacuum are used. At 50 °C the vapour pressure of styrene is less than 50 mBar, so 95% of the air could be evacuated from the tool before the polyester resin began to boil. This level of vacuum would normally result in a great improvement in quality and would help to avoid air entrapment in more complex tools. For the highest levels of tool complexity it can become essentially impossible to define in and out gate positions and process windows that ensure complete wet-out without any possibility of air entrapment, especially when preform positional tolerances are considered. In these cases the dominant process parameter can be the level of vacuum and absolute pressure levels of a few mBar are not uncommon. In such cases the use of polyester resins may not be possible and care would have to be taken that reinforcements were thoroughly dried. It has been suggested that hot tools could be completely evacuated of air then charged with styrene at about 100 mBar to permit the use of polyester resins for even the most complex tools without any consideration of out gates positions (or indeed any out gates). It is not known whether this has proven possible and it is usual to utilize epoxy resins when a high vacuum level is required.

The capabilities of the injection equipment also need to be taken into account. Many mix and meter pumps are designed for room temperature operation. While it is perfectly possible to mount them in a heated cabinet there can, for example, be problems with seals at a high temperature. In this case either re-equipment is needed or a resin must be chosen that has the right viscosity characteristics within the operational envelope of the pumps. The same is true of curing temperature requirements set by available process equipment or tool material temperature limits. For complex components a lower cure temperature is preferred to minimize cure and thermal stresses.

With respect to thermal and cure stresses low shrinkage systems are best, as is a low-expansion coefficient, a low Poisson's ratio, and a Tg above the cure temperature (because the expansion coefficient is much higher above Tg and in complex geometries complete stress relaxation may not always be possible due to the presence of a full 3D constraint). Another factor that leads to high thermal stresses in resin rich zones formed in complex geometries is the resin's modulus. It is usual to aim for a high modulus, but if this is done stresses will be maximized and a lower modulus could be preferred (so long as mechanical properties are not compromised). High toughness, strain to failure and strength will also help to limit the likelihood of cracking in resin rich zones. These factors also control the resin's resistance to thermally induced microcracking. It would also be useful to know the behaviour of resins after gelation, especially in terms of the relationships between the development of resin strength as cure progresses and shrinkage-induced strains in constrained resin richness, but data seems to be lacking. RTM resins vary widely in their resistance to microcracking and resin rich zone cracking, unfortunately data on the issues raised above are seldom forthcoming from the resin manufacturers. Older and more brittle epoxies, and bismaleimides, have the poorest behaviour in these respects and newer systems seem to be much more resistant to thermal effects. Flight hardware has been made with old and brittle systems that exhibit high levels of microcracking, without any apparent effect on component reliability. In general it seems prudent to utilize the more recent formulations that are less sensitive to temperature changes. Having said all of the above the state of knowledge about the causes and correction of cracking and delamination in thick section composites is still far from complete.

The use of resins in which a great deal of shrinkage occurs before gelation can lead to intertow voidage, especially in rigid tools, as noted earlier. While it is obviously better to avoid these, I know of no tests to measure this property in a standardized way. Mercury bulb dilatometry results have been reported,[14] and this seems to be the best option currently available. All that can be said here is that if such problems become apparent it is worthwhile to try changing the resin in order to try and distinguish between flow irregularity induced voidage and shrinkage induced voidage. Keeping the resin under pressure during cure would also be of benefit in such cases, so long as resin could be fed back into the tool cavity to offset the shrinkage. In practice this means that it would be best if resin cure commenced at the centre of the tool and moved outwards towards the resin feeds.

Resin selection for RTM is much more than merely ensuring that viscosity is low and working time is high enough, especially when geometry is complex. All factors, from the application that the component is intended for to the details of the production engineering route to be followed need to be taken into account.

2.2.2 Polyester and vinyl ester resins

The range of these resins is enormous and the number usable in RTM only slightly less so. Equally, a very large number of suppliers exist from very large companies offering a full range of resins and good technical support to small companies offering low-cost products over a narrow range. In the context of advanced applications it is probably more common to utilize epoxies, but there is no reason why the advantages of these resin types in terms of cost and cure rate cannot be exploited. It must be said that such resins lack the durability of many epoxies and that good property databases tend to be lacking, as do allowable properties. For non-aerospace applications it would be well worth while considering this class of resins, provided that the supplier can provide an acceptable level of reliability and technical support. The range of possibilities is far too large to make any recommendations beyond the obvious one that the choice of supplier may be as important as that of a specific system.

2.2.3 Epoxy resin systems

The majority of the major suppliers of epoxy resin systems are now producing resins that are specifically designed for use in RTM. In addition, many grades of resin that are not specifically formulated for RTM are usable in the process. These include resins formulated for wet filament winding, as many of the process requirements are similar. References 15, 16 and 17 describe resin developments in some of the major suppliers. One development of note is the trend towards the supply of premixed resins that are supplied as single component systems.[17] These have great advantages in that control of mix ratio is in the hands of the supplier. This has the effect of simplifying both the machinery requirements for the injection phase and the quality control and record-keeping requirements for the most critical components. The suppliers have also carried out work on dispensing systems suitable for use with their materials, so that one supplier can be approached for both materials and equipment selection. Some epoxy resins recommended for RTM contain MDA, a material that is toxic and requires close monitoring of

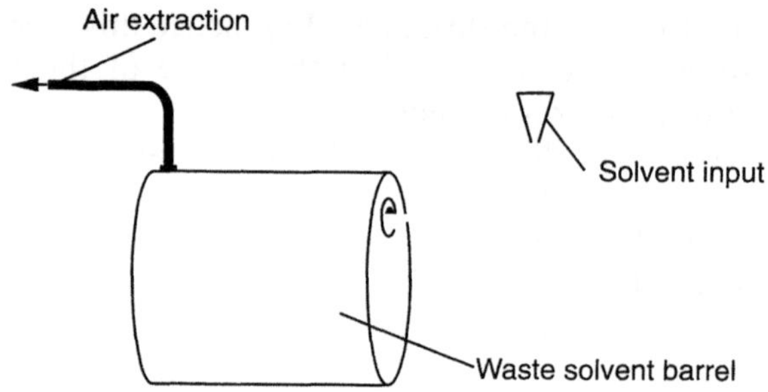

Figure 2.8 Close coupled equipment is needed for solvent flushing of injection lines with some systems.

operators. In principle RTM permits reduced contact with resins by the operators. I have seen many examples where, due to mixing requirements or poor equipment choice, there is a great deal of resin contamination of tools and production areas. If MDA containing resins are to be used it is axiomatic that contact with the resins be avoided and only the very best production practices must be followed. As an example, if solvent flushing is used, the flushing equipment should be close-coupled to a waste solvent barrel to avoid splashes of contaminated solvent (see Figure 2.8).

As the requirements for the resin vary so much from part to part it is impossible to single out one resin or one supplier as being the best for RTM. In the context of advanced applications it is preferable to use resins from one of the major suppliers that can supply the required levels of traceability and technical support, especially whether allowable properties have been generated for the systems required and the scale of the available database. Absolutely critical to the question of traceability is that the supplier must be able to unequivocally guarantee that there will be no changes to the formulation of the resin or to any of its constituent parts. Ideally, the chosen resin supplier should manufacture all the resin ingredients in house to ensure that it can make the required guarantees. For many components cheaper and simpler systems are adequate. Improvements have been made in terms of both ease of use and general toughness, but many of these systems are considerably more costly than the older systems. As always the total suite of costs and benefits must be carefully weighed before deciding on any particular system.

2.2.4 Phenolic resins

Phenolic resins are normally used where a combination of low cost and fire resistance is demanded. Resins that employ resol chemistry and acid catalysis are the most common, although some advances are being made in reducing the acidity of the catalysts. Conventional acid catalysts are very corrosive, both to equipment and personnel. Equipment can and should be made fully acid resistant. Personnel cannot be made acid resistant and equipment and operating practices must reflect this. When phenolics cure, water is evolved, in addition many resins use water to control the viscosity of the resin. From my own observations, the cure of acid catalysed phenolics proceeds by a slow thickening until patches of 'milkiness' appear on the surface of the resin. At this point the milkiness rapidly spreads through the resin, accompanied by a great increase in viscosity and heat generation. The 'milkiness' is associated with a phase change within the curing resin, when the resin can no longer hold the evolving water in solution. The water seems to be ejected on a very fine scale, with the resin viscosity rapidly rising before the distribution of water droplets has any time to coarsen. Thus the structure of the cured resin has the appearance of a dense, fine cell foam, with initially water-filled voids a few microns in diameter. The fact that the water is evolved in this way, rather than as larger droplets, permits RTM processing of these acid catalysed phenolics, so long as peak exotherm is not so high as to lead to the water boiling. Considering the structure of the cured resins the mechanical properties are, perhaps, surprisingly good, although the resins are generally brittle. Phenolic resins sometimes show problems of compatibility with reinforcements, in that wet-out may not occur naturally or sufficiently rapidly: leading to quality problems. This is at least as much a problem with the surface treatments or coupling agents used on the fibres as it is with the resin, and combinations should be carefully checked before being specified. There are fewer suppliers of phenolic resins than of polyesters or epoxies. In view of the potential processing problems with phenolic resins the choice of the right supplier that can provide the maximum level of technical support may be vital.

2.2.5 Bismaleimide resins

Bismaleimides offer the heat and fire resistance of phenolics, coupled with the ease of processing, convenience and (almost) the level of properties of epoxies: at a price higher than either. Early BMIs that had viscosities suitable for RTM tended to be extremely brittle, greatly limiting the potential geometrical complexity of components if extensive cracking was to be avoided. Much improved BMIs are now available, although less work has been done than on epoxies and the number of suppliers is lower. Even the best current RTM processable BMIs are less tough than the best available epoxy resins.[18, 19]

Other resin systems suitable for RTM have been developed or are under development.[20, 21] By far the bulk of the market is taken up by those noted above, roughly in the order noted.

2.3 BINDERS

Binders are used in many techniques that permit off-line shaping of reinforcements into preforms that are robust and handleable, to simplify mould loading and improve quality. Binders can be broken down into four types.

1. Tackifiers and sticky web materials, these are mostly used for 'hand layup'.
2. Carrier removal is used with emulsion binders where the carrier is generally water. It can also be used with binders that can be softened by solvent sprays.
3. Thermally softening binders, these can be in the form of powders or fibril webs (a lighter form of those used for interlining in garment production). The binder is solid at room temperature, it can be a true thermoplastic or based on resin chemistry.[16, 22]
4. Curing binders can be of more or less any form or chemistry and thermal and UV cure have been reported.[23]

Within this range of binding options some general comments can be made.

2.3.1 Basic requirements

The function of the binder is to create a handleable preform. It does this by creating local bonds both within each ply of reinforcement and between the plies. These bonds must be formed while the reinforcement is held into the required geometry and be adequate to hold that shape when forming forces are removed. The volume of binder holding together the preform is generally small, a few per cent of the reinforcement weight.[24] For this reason the binder must be adequately strong and tough and it is common to use thermoplastic species as binders. If the binder is adequately strong after binding some means must be found to soften the binder prior to forming. This can be done by heat or partial dissolution for multiple use binders (to distinguish them from curing binders which are only capable of one binding cycle). The binder viscosity must not fall too low during processing or the binder may be wicked into the reinforcement and give poor interply bonding. Equally, too high a viscosity will lead to poor bonding; a minimum viscosity in the hundreds of poise range seems to be about right, so long as the cycle time is not excessive. It is probably easier to control the process of thermal softening than partial dissolution in organic solvents, and the second option has obvious health and safety implications. For thermal softening, the melt temperature needs to be high enough to avoid relaxation at room temperature and low enough to avoid damaging the reinforcement or its surface finishes. A temperature in the low 100s °C seems to be in the correct range, although lower or higher temperatures have been used.

2.3.2 Binder types

Powder binders are commonly used. Adequate results can be obtained, with care, by manual application from a shaker. For the maximum uniformity, powder coating units are required. The use of such machines would be expected to be mandatory for aerospace applications. The powder is generally affixed to the reinforcement by radiant heating, but other options would be possible dependent on binder chemistry. Powder bound reinforcement types are commercially available. The binder tends to sit on the surface of the cloth; as such it performs very well in binding together multiple layers. It performs less well in binding single layers of reinforcement into shape unless higher binder proportions are used. Powder bound cloth can be bound on one or both sides. Binding on both sides eliminates problems with errors in preform assembly (i.e. two unbound faces opposite each other) but requires two passes through the coating line and may lead to problems with sticking to preforming tools. In general, single-sided bound cloth is used.

Fibril binders are made by passing the molten binder material through a fibrilator gun that produces a fine web of fibrils of binder, either directly on the reinforcement or onto a paper backing sheet, from where it can be ironed onto the cloth. Rolls of fibril binder are commercially available. They are useful for small jobs where the expense of using a powder coater is not justifiable but binder uniformity must still be well controlled. At a constant weight/unit area fibril binders are more effective than powder binders at binding single layers and less effective at binding multiple layers.

Liquid binders can be emulsions or solutions. Emulsion binders can be held in suspension in a liquid and sprayed onto the reinforcement just prior to binding. The liquid is then removed during the preforming process. In one variant a rubber-faced vacuum box is used to form the reinforcement and dry the preform at room temperature. In this case the preform tends to be built up one ply at a time. If a solution of thermoplastic is used the sheets of reinforcement can be pressed between heated tools to drive off the solvent. In this case the binder can permit each ply to be pressed to final dimensions outside the mould tool, this can greatly assist in reducing potential problems with mould loading. Higher levels of binder seem to be used in this process, restricting resin flow rates

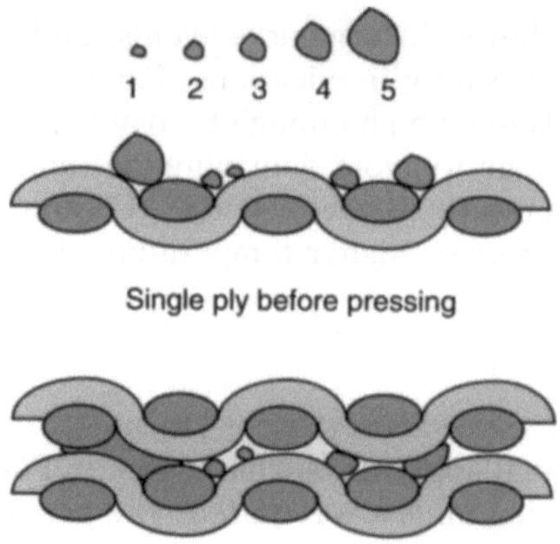

Figure 2.9 The particle size of powder binder can have an influence on binder performance. Five particle sizes are shown. Sizes 1 and 2 are rather too small and will contribute little to binding. Sizes 3 and 4 are probably in the right size range and size 5 may be a little too large.

during injection.[25] A solution of a resin that is solid at room temperature can be used in the same way, with less influence on resin flow. These sorts of binders perform better than other types in holding single plies into shape, but perform less well with multiple plies.

2.3.3 Powder binder dimensions

Figure 2.9 shows a cross section through a bound cloth with different particle sizes of binder. The very smallest sizes (1) may be oversoftened by radiant heat or solvent application and be absorbed into the cloth to have little binding effect. Even if they do not become oversoftened the smaller particles (1 and 2) can fall into the spaces in the reinforcement, again contributing little to the binding between plies. Larger sizes (3 and 4) should give good binding effects. The largest binder particles (5) can produce problems. If they are not fully softened they can tend to hold the plies apart or deform the reinforcement structure. If the binder is insoluble in the resin an oversize particle might act as a weakness, equally an oversize particle might not dissolve completely in the matrix resin. Figure 2.9 shows an optimum binder particle size of about the tow thickness. In general this seems to be appropriate, although details of binder rheology or compatibility with the matrix might make other choices of particle size more appropriate. For any particular reinforcement/binder combination some experiment may be required to find the best particle size. A good starting point for experiments would be a particle size equivalent to, or slightly larger than, the finished tow thickness with as narrow a size range as possible.

2.3.4 Binder/resin interactions

The binder is going to form a permanent part of the composite structure, its compatibility with and interactions with the matrix are therefore of great importance. Compatibility essentially rests with an understanding of both the resin's and the binder's chemistry and must be considered on a case by case basis. Some general points can be made here. The first is that the ultimate arbiter of compatibility for highly stressed components is the measurement of allowable mechanical properties, in the use environment, using material made in exactly the same process as will be used for production. Thus any binder to be used in production **must** be part of the laminates made for allowables testing and the overall process **must** be as intended for production. Binders may be soluble or insoluble, they may be liquid or solid at the injection temperature and or cure temperature. If the binders are liquid at the injection temperature they may wick into the tow if viscosity is low, or mix/dissolve in the resin if viscosity is higher. This mixing may increase resin flow front viscosity and may lead to preferential binder concentration in some areas of the moulding. The latter effect is limited (for high-volume fraction, organized reinforcement) by the observation that the resin in the tow moves very little after it is wicked into the tow. If the binder is solid at the injection temperature, but still soluble, the same effects can occur. Another approach is to use a binder that is barely soluble at the injection temperature but which becomes soluble between the injection and cure temperatures. Hot-stage microscopy can be carried out to determine relevant temperatures for injection and dissolution of the binder. In such experiments I have observed a very clear effect of increasing dissolution rate at very small particle sizes, which is another reason for maintaining a tight particle size distribution. For such, controlled dissolution, binders the effective formulation of the resin in the tows (largely free of dissolved binder) and between the tows (high concentration of dissolved binder) may be rather different. This is one reason for insisting that allowables be measured at exactly the process conditions to be used in production. It should be noted that if this approach is followed there are additional constraints on the possible process windows, both for injection and cure stages. These considerations can be avoided by using insoluble binders that are solid throughout the processing and use of the components. The difficulty here is that if the binder is to be softened enough to bind at temperatures below those that might damage the reinforcement it is quite possible that the binder may be beginning to soften within the operational temperature range of the components made. As before the final arbiter of acceptability is the component requirements and allowable properties as measured on the reinforcement + resin + binder system. Curing systems can be used to overcome all the problems noted above, but they introduce their own limitations with bonding between cured binder and resin and cannot be used for the multiple bonding cycles that are sometimes needed for complex preforms.

The most commonly used binder types for powder or fibril applications have been based on thermoplastic polyesters. These materials have very good binding properties, but their compatibility with epoxies is at first sight questionable. Despite this, such materials have been used in highly loaded components without evidence of binder related problems, as measured in allowables programmes or in service. In recent years, attempts have been made to develop improved binders with some success.[26] To date most work on binder chemistry has essentially been negative. That is to say that the aim has been to minimize losses in properties due to the binder and understand and control what happens to the binder during processing. There are indications that this is unnecessarily pessimistic. Work on interleaving laminae with tough layers of thermoplastics,[27] indicates that great improvements in impact toughness and reductions in crack propagation can be achieved. A suitably formulated binder could be used to achieve the same effects, I have certainly seen preliminary results that show great improvements for Compression After Impact results from following this approach. Current RTM epoxies lack the highest levels of toughness, largely because they must be formulated for low melt viscosity. Optimum formulation of resin and binder together offers perhaps the only route by which RTM laminate properties can be increased to the levels found in the best prepreg epoxies, without sacrificing processability.

2.4 CORE MATERIALS FOR RTM

There are various reasons for utilizing core materials in RTM. These include: conventional sandwich panel strengthening/stiffening; local reinforcement for fastener locations; the creation of bonding lands; thermal insulation; easing the manufacture of complex geometries by eliminating undercuts; provision of a compliant core to control reinforcement compaction pressure etc. In order to prevent the core from filling with resin during injection it is usual to specify closed cell core constructions. In general this rules out the use of honeycomb cores. These can be used if a sealing film is bonded to the core prior to resin injection. I have manufactured small, flat test panels in which the core was encased on both sides by a film adhesive and suitable film, with reinforcement placed on both sides of the assembly in the tool. In this case the film adhesive cure was included in the process cycle such that cure was complete prior to injection of resin. The resultant panels appeared to be of similar strength to more conventionally moulded honeycomb sandwich panels.

The traditional core material in general industry used to be end grain balsa, but this has largely been replaced by synthetic foams of various chemistries. Cast to shape polyurethanes have been widely used, for example in aircraft propellers by Dowty and in special purpose vehicles.[28] As cast these materials have a skin that limits the bond strength to the injected resin. This must be removed by scraping or shot blasting prior to use, which can reveal defects that need to be repaired. As an alternative, Dowty is believed to use a proprietary material to line the foam tool. This bonds well on one face to the foam and also bonds well on the other face to the injected resin. In general, cast to shape polyurethane foams lack the strength of some types of block foams such as PVC and polymethacrylimide (PMI) foams.

These block foams are widely used in advanced RTM. The PMI foam sold as Rohacell has excellent properties and is finding application in operational aircraft components.[29] PVC foam is the standard material in the marine industry. These foams have to be cut to shape, although some measure of thermoforming is possible with some types. If foam inserts are to be used for local reinforcements for fasteners it is usual to use the denser and stronger syntactic foams, generally as precured inserts.

REFERENCES

1. Turner, M. R., Rudd, C. D., Long, A. C., Middleton, V. and McGeehin, P. (1995) *J. Advanced Composite Letters* **4**, 121–4.
2. Summerscales, J. (1993) A model for the effect of fibre clustering on the flow rate in resin transfer moulding, *Composites Manufacturing* **4**, 27–31.
3. Basford, D., Griffin, P., Grove, S. and Summerscales, J. (1995) Relationship between mechanical performance and microstructure in composites fabricated with flow-enhancing fabrics, *Composites* **26**, 675–9.
4. Heins, G. and Jackson, P. (1987) Novel reinforcing fabrics for high performance composites, *Proc. 42nd Annual Conf. Composites Institute*, SPI, Feb., Session 7B.
5. Hogg, P. J., Ahmadnia, A. and Guild, F. (1993) The mechanical properties of non-crimped fabric based composites, *Composites* **24**(8), 423–32.
6. Raz, S. (1988) *The Karl Meyer Guide to Technical Textiles*, Obertshausen: Karl Meyer Textilmaschinefabrik.
7. Parat, I., Greenwood, K. and Li, Z. (1996) CAD/CAM of 3D woven structures (preforms) for fibre reinforced composites, *Composites*, Part A, 27A, 111–17.

8. Bruno, P., Keith, D. and Vicario, A. (1986) Automatically woven 3D composite structures, *SAMPE Quarterly* **17**(4), 10–17.
9. Brown, R. (1985) Through the thickness braiding technology, *Proc. 30th Annual SAMPE Symposium*, March, 1509–18.
10. Olsen, N. (1986) Advanced manufacturing technology for structural aircraft/aerospace components, *Proc. 31st Int. SAMPE Symposium*, 387–93.
11. Verpoest, I., Wevers, M. and De Meester, P. (1989) 2.5D and 3D fabrics for delamination resistant composite laminates and sandwich structures, *SAMPE Journal* **25**(3), 51–6.
12. Wang, C. Z. *et al.* (1989) Fracture toughness and failure modes of interleaved fibre composites. Economic comparison of advanced composite fabrication technologies, *Proc. 34th SAMPE Symposium*, 1497–1506.
13. Walker, N. (1987) Veils, mats and tissues for non-structural applications, *Proc. ICCM6/ECCM2*, July, London: Elsevier Applied Science, 5. 547–56.
14. Yates, B., McCalla, B., Phillips, L., Kingston-Lee, D. and Rogers, K. (1979) The thermal expansion of carbon fibre reinforced plastics. Part 5. The influence of matrix curing characteristics, *J. Mat. Sci.* **14**, 1207–17.
15. Stark, E. *et al.* (1987) New non-MDA epoxy resin systems for RTM and filament winding, *32nd International SAMPE Symposium*, April, 1092–1103.
16. Puckett, P. and White, W. (1990) Thermoset resin systems and manufacturing technology for RTM. In *Resin Transfer Moulding for the Aerospace Industry*, Los Angeles: SME, 6–7 March.
17. 3M Datasheet, PR 500 RTM resin system.
18. Stenzenberger, H. D. *et al.* (1989) Advanced composites processing with bismaleimide resins. Materials and processing – move into the 90s, *SAMPE European Chapter Conference*, July, 277–92.
19. Stark, E., Breitigam, W., Farris, R., Davis, D. and Stenzenberger, H. (1990) RTM of high performance resins, *Proc. 35th International SAMPE Symposium*, April, 782–94.
20. Okamoto, Y., Klemarczyk, P. and Levandaski, S. (1993) Novel vinyl ether thermosetting resins, *Polymer* **34**(4), 691–5.
21. Stockton, J. (1989) Structural resin transfer molding of high temperature composites, *34th International SAMPE Symposium*, May, 1032–40.
22. 3M PT500 binder datasheet.
23. Horn, S., Buckley, D. and Seroogy, K. (1990) High volume, highly automated preform process for RTM and SRIM, *45th Annual Conf. Composites Institute*, SPI, Feb., Session 9-C.
24. Hansen, R. S. (1990) RTM processing and applications. In *Resin Transfer Moulding for the Aerospace Industry*, Los Angeles: SME, 6–7 March.
25. Jones, W. and Johnson, J. (1980) A resin injection technique for the fabrication of aero-engine composite components, *Proc. Syposium: Fabrication Techniques for Advanced Reinforced Plastics*, April, Salford: IPC Science and Technology, 40–7.
26. Kittelson, J. L. and Hacket, S. C. (1994) Tackifier/resin compatibility is essential for aerospace grade RTM, *Proc. 39th International SAMPE Symposium*, April, 83–96.

27. Masters, J. E., Courter, J. L. and Evans, R. E. (1986) Impact fracture and failure suppression using interleafed composites, *31st International SAMPE Symposium*, April, 844–58.
28. McCarthy, R., Haines, G. and Newley, R. (1994) Polymer composite applications to aerospace equipment, *Composites Manufacturing* **5**(2), 83–93.
29. Akay, M. and Hanna, R. (1990) A comparison of honeycomb core and foam core carbon/epoxy sandwich panels, *Composites* **21**(4), 325–31.

3 Reinforcement manipulation and preforming

3.1 INTRODUCTION

The function of preforming approaches is to assemble reinforcements into the required geometry in a reproducible manner, with the minimum cost and with the possibility for inspection and quality assurance steps. Three general approaches seem to have been developed, which could be used alone or in combination for any specific component.

The first approach is to weave, knit, or braid net shape or near net shape preforms. Dry filament winding would also fall into this group, although the geometries possible are more limited. Despite the advances that have been made these preforms are somewhat limited in geometrical complexity and there seems to be a steep trade-off between speed of production and performance. For example, complex, plain knitted shapes can be made rapidly but at low Vf% and performance. On the other hand complex braiding can produce high-performance preforms, but only slowly. There is another consideration here with respect to production engineering constraints. If production rate is to be increased, the generation of preforms via capital-intensive machinery that must be specially built, or has a long lead time, may be a limiting factor. In general the whole of the production route should be designed for the eventual production volume and if changes in production volume are expected it is necessary to look very carefully at the cost and time factors associated with these changes.

The second approach is to sew together the reinforcements to be used into a flexible, or semi-flexible, reinforcement pack for insertion into the tool. This approach has mostly been developed in the USA.[1] This may be the best approach for very large structures, but some problems are obvious. Handling large flexible assemblies will be difficult, very complex geometries would be expected to be hard to achieve, and achieving reliable tolerances with what are in effect very sophisticated dressmaking techniques may be problematical. Despite these potential problems, very good results have been achieved using very dense stitch patterns. If foam cores are used the reinforcement can be stitched or stapled directly to the foam, which will help to stabilize the preform somewhat. The amount of stitching used will also affect the bulk factor of the reinforcement, and thus the ease with which it can be loaded into the tool without dangers of fibre pinching or movement during mould closure. If the layers are simply stitched lightly to hold them together little reduction in bulk factor will be seen. If a more dense stitching is used a reduction in bulk factor and more reliable mould loading should be experienced. In this case some improvements in damage propagation may also be seen. As always there will be a trade-off between preform production costs, mould loading costs and costs elsewhere in the manufacturing cycle, with the optimum solution only discoverable on a case-by-case basis.

The last approach utilizes the binders discussed earlier. These can be used to lay up in a ply-by-ply manner analogous to the use of prepreg. This can be rather difficult with thermally softening binders, and tackifiers or sticky web materials can ease the production of preforms made in this way. This approach has the benefit of permitting operators trained in the use of prepreg to prepare preforms without great retraining, but overall process economics are unlikely to be greatly improved compared to the use of prepreg. In some cases, due to the geometry of the part, no other techniques are possible and this hand layup approach must be used. In one case of a complex component studied by the author the only layup option was the use of a thermally softening binder, based on a BMI resin, in a ply-by-ply manner. To enable the layup to be eased some simple tools were designed and built. These consisted of a temperature controlled soldering iron, with a variety of tips conforming to different areas of the geometry. Small blocks of rubber cast to the required geometries of different parts of the tool were also used. With a little practice it was found that these simple tools enabled a quick, efficient and high-quality layup. The final components were of a very high quality when made by RTM using these preforms, whereas the same components made by prepreg/autoclave techniques could not be made at acceptable quality; essentially because of the geometrical complexity.

In general, bound material preforming techniques use forming tools that generate preformed shapes that can be assembled together to create complex geometries. This assembly of complex shapes from simpler mechanically formed preforms leads to an approach to the mechanization of composites production that eliminates many of the geometrical limitations inherent in other automation techniques. To understand the implications of this approach an analogy can be taken with the production line assembly of automotive body parts. Each sub-system of the body shell is broken down into a series of sub-components. These are formed on simple tooling with low labour contents and assembled into larger structures by spot welding or bonding. It is unlikely that this process would ever produce the most weight-optimized structures, but its overall cost-effectiveness is undoubted; at least at high production volumes. The reason for this last caveat is that metal stamping tools and presses are very costly and this can dominate the overall production economics unless volumes are very high. By contrast, most preforming tools can be very lightly designed as forming loads are often almost insignificant, leading to low-cost suites of preforming tools. In the automotive case the overall performance of the parts, and to some extent their geometry, is controlled by the approach chosen of bonding or spot-welding sub-assemblies. From this analogy it is to be expected that the adoption of the assembled preform route will require changes to be made in the way that components are designed, to reflect the characteristics of the processes involved.

The baseline process utilizes bound reinforcement that is thermally softened and formed in cold tools, before being trimmed to size and assembled either into the tool or into a larger preform assembly. Other possible processes, such as the use of curing binders, will have similar characteristics and only the baseline process will be discussed here. When preforms are made from a single ply of reinforcement all deformation essentially takes place in the plane. To prevent the relaxation of this deformation requires a higher level of binder than if multiple layers are formed and it is normal to preform two or more plies at once. When two or more plies are formed, shearing is required between the plies, in addition to within each ply. Binding the layers together to prevent the relaxation of this inter-ply shear acts to stabilize the shape of the preform. Experience has shown that good, handleable preforms can be made more easily with several plies than with single plies. For this reason, and the overall improvements to process economics that arise from handling several plies at a time, the baseline process might be further defined as the formation and assembly of multi-ply preforms.

To attempt an understanding of the ways in which the adoption of the baseline process will affect design when a ply-by-ply approach based on prepreg is replaced by the assembly of preformed ply blocks various factors need to be examined. These include the deformation characteristics of reinforcements, preform tooling and machinery requirements, quality control requirements etc. Before considering these factors it should be emphasized that this approach can be utilized in concert with other preforming techniques. The author has been involved in projects where braid has been used to form some elements of the final preform and stamped preforms and dry filament winding have been used for others, with final assembly by stitching the details together.

3.2 DEFORMATION MODES OF COMPOSITE REINFORCEMENTS

When a sheet or sheets of reinforcement is press formed three results can be generated. A draped sheet can be formed that covers the forming tool exactly. The material's forming limits can be exceeded leading to tearing or other damage to the reinforcement. If forming limits are exceeded or control of the forming process is lost then folds or wrinkles can be formed. These defect types must be avoided and the starting point for this is to gain some understanding of the deformation modes and limits in the various types of reinforcement. Having said this it must be admitted that such an understanding is far from fully developed and that an element of trial and error is generally present in the design and development of most preforming tools.

For an arbitrary layup of sheets of reinforcement to be deformed into an arbitrary shape, extension is required along and transverse to the sheet axes and in plane and out of plane shears are required. During this global deformation the local fibre trajectories should remain smooth and unkinked.

Different types of reinforcement behave in very different ways,[2] In general two types of load/extension behaviour can be identified (Figure 3.1).

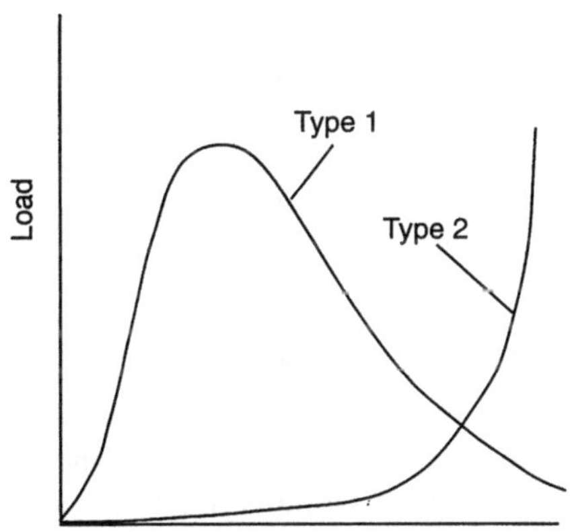

Figure 3.1 Different types of reinforcement show different load/extension behaviour when subjected to forming forces such as those seen in preforming. The type of behaviour influences the preforming approach. Type 1 materials behave unstably after some deformation and may develop defects. These materials include random mats and bound UD material transverse to the fibres. Type 2 materials behave in a more predictable manner. These materials include woven cloths and plain knits. Some materials such as warp/weft insert knitted materials tend towards type 2 in large sheets and type 1 in small pieces.

Type 1 materials show a rise in load with extension up to a peak, beyond which the load starts to fall as defects are formed and propagated. Deformation is irreversible. The forming limit can be associated with the peak in the curve, although this can often be shifted upwards by constraint such as might be supplied by rubber backing sheets. Materials of this form include random mats and aligned short fibre mats in longitudinal and transverse tension and shear and bound UD materials in transverse tension and shear. Type 2 materials show a slow rise in load up to some point where fibre–fibre interactions lead to a rapid rise such that forming forces become excessive. Deformation is reversible. In principle if deformation was continued these materials would also show a load peak as the material failed, the distinction between type 1 and type 2 is fundamentally between irreversible and reversible deformations. This group includes woven cloths in shear and plain knits in any loading case. For small sheets the above distinctions break down somewhat. If a small piece of biaxial woven cloth is deformed, deformation will initially proceed by plain shear or scissoring between warp and weft. As loads rise some slip may occur in which tows are drawn through the structure of the cloth leading to behaviour between the two basic cases.

Deformation limits may be measured for the materials required. In practice lower limits may apply if control of the forming process is lost. Guides are sometimes used to limit the movements of the reinforcement as it is drawn into the preforming tool. Whenever a reinforcement is deformed there will be changes in the structure of the reinforcement. These will be changes in thickness and fibre directions and are dependent on reinforcement type and deformation mode. Excessive changes in ply thickness can be a particular problem in rigid tool RTM as they can lead to high or low Vf% areas which greatly distort resin flow front shapes. Specific reinforcement types are considered below.

3.2.1 UD bound mats (with the binder in the softened state)

For continuous fibre mats no deformation is possible in the fibre direction, some deformation would be possible for aligned short fibre mats.[3, 4] Transverse deformation is possible (type 1) up to the onset of transverse splitting. This is likely to be at a low elongation for continuous fibre mats. Shear is possible but forming limits would be expected to be low in the absence of constraint. In principle only one free edge is required for forming into 3D geometries (Figure 3.2); in practice, forming limits would be low.

Bonded +/– X angle ply layups will initially deform in a similar way to woven cloth, but separation between the plies will occur at some point.

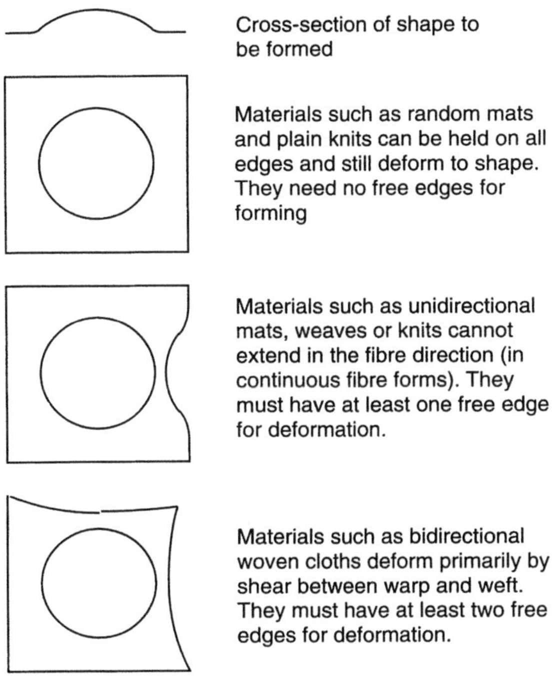

Figure 3.2 The differing deformation mechanisms give rise to differences in the deformed shape of a square piece of reinforcement deformed over a common form.

3.2.2 Woven and knitted unidirectional

No deformation is possible in the fibre direction. Transverse extension (type 2) may be possible depending on the tightness of the weave or knit fibres. Knitted variants tend to be more transversely deformable than woven. In principle thickness would be proportional to deformation, in practice, especially for knits, the tows will move apart at a more or less constant thickness. Shear is possible and will be much more stable than for bound forms of UD. As for bound UD only one free edge is required although deformation limits would be expected to be higher.

3.2.3 Woven and knitted biaxial

No deformation is possible in the warp or weft directions. Shear is possible from the initial +/–45 to some new angle (type 2). For pieces of cloth large compared to the weave dimensions the cloth will deform in a predictable way as a pin-jointed net up to large deformations.[5] In small pieces of cloth some slippage of tows through the cloth can occur, leading

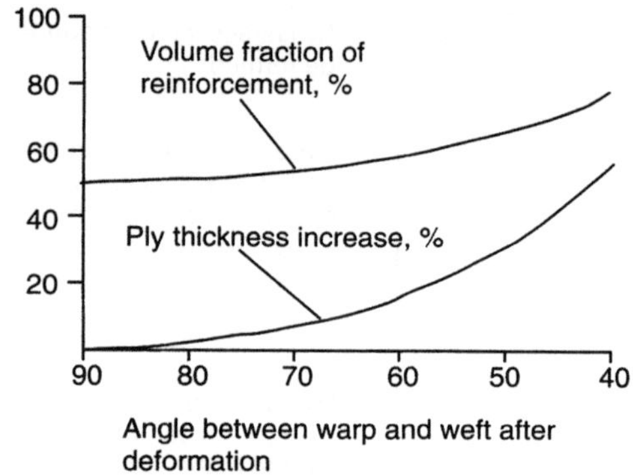

Figure 3.3 When woven cloth is deformed the cloth thickness changes in a predictable manner. Cloth preforms will be thickest in regions of the highest deformation. If the increased thickness is pressed down to a constant thickness the local Vf% can become excessive, resulting in low permeability areas. Minimum warp/weft angles depend on cloth type, usually lying between 40 and 50 degrees.

to less predictable behaviour. Bound cloth is rather less likely to slip in this way than unbound cloth. Because only one deformation mode is available two free edges are required for deformations out of the plane, making the material less drapeable in theory than bound UD. In practice the shear mechanism with its high deformation limit can give rise to very deformable materials. Cloths have been specifically designed for the maximum drapeability. While the cloth deforms as a pin-jointed net its deformation is predictable and software packages are available to define both the final fibre angles and the developed flat shape corresponding to the deformed shape,[6] although the software seems to be limited in the geometries it can handle. The thickness of the cloth is also predictable (see Figure 3.3).

At the limits of deformation these changes in thickness can be very great and should be accounted for in component and/or tool design. Knitted biaxial cloths behave in a similar way, although the tendency for tows to be drawn through the material is higher due to the lack of crimp. Some of the knit patterns that are used to hold the warp and weft tows can lead to fibre wrinkling when the material is deformed. This should be checked on small samples of cloth prior to use. Cloth is very useful in hand layup precisely because its reversible load extension behaviour permits repeated attempts to drape and position a ply. In mechanical forming of highly deformed shapes high forces can be generated, leading perhaps to tool wear or fibre damage. In some cases double-acting tools are used to overcome this and similar problems. Lastly, it is not axiomatic that cloth starts the deformation process at +/–45, sheets of cloth can be predistorted to other angles prior to forming and this can occasionally be a useful option.[7]

3.2.4 Multiaxial knitted materials

Theoretically these materials should be inextensible in all directions due to their triangulated structures. In practice tows can be drawn through the knit fibres to give good deformability (type 1, at least in so far as the deformation is essentially irreversible). This mechanism is poorly understood and predictive tools for final angles and thicknesses are lacking. Preform development would, therefore, have to be on a trial and error basis. It should be noted that when the tows are drawn through the knit fibres they are held in position by them such that when the forming forces are removed the formed shape does not immediately spring back to the undeformed shape. This offers the potential for producing shaped preforms without the necessity for any separate binder, although preforms made in this way seem to be less crisp than those made, for example, with bound cloth and less semi-permanent reduction in bulk factor would be expected.

3.2.5 Plain knitted materials

These materials are deformable in all directions, by a process of straightening out the loops in the knit until the fibres are effectively straight and deformation ceases (type 2). They can be highly deformable and locally embossable. They tend to be used primarily as surfacing materials, although they have some advantages over random mats in terms of fibre continuity, uniformity of weight/unit area and freedom from deformation induced defects.

3.2.6 Random in the plane mat materials

Deformation is possible in all directions (type 1). Embossing is possible, i.e. no free edges are required for deformation. The continuous fibre forms of these materials would be expected to have higher deformation limits than short fibre forms. Deformation will be limited by localized thinning leading to non-uniform properties. Needling and stitching can be used to give better uniformity of deformation in these materials. At best these materials can be highly deformable, but properties are low.

3.2.7 Complex woven sections, e.g. 'T' or 'H' sections

Some shearing between warp and weft is generally possible, permitting single curvatures or twisting, but these materials are very limited in double curvature capabilities.

For all the above materials, when used in multi-ply layups the out of plane shear is at least as important in the prevention of wrinkles as the in-plane deformations. In general the binder must be fully softened throughout the reinforcement stack to permit good forming, which represents a process limit on the rate at which preforms can be produced. If thermal soak times are excessive for a thick stack it can be advantageous to split the stack into thinner stacks, depending on the type of heating used and the relative times for heating and forming. Forming in cold tools is usually very rapid as the stack must be formed before the preform cools excessively. Little work seems to have been done on these issues but Potter (1983)[8] discusses similar problems in thermoplastic matrix composite forming.

Any and all of the above noted reinforcement forms can be used in preforming. Even those forms with very limited deformability can be of use when preforms are essentially of single curvatures and in some cases the art of curved line folding can be used to generate 3D geometries from essentially undeformable materials (see Figure 3.4).

In practice, when designing preform tools, it is more usual to rely on experience of the practical limits on deformability than on the available and limited models. For example Figure 3.5 shows that it is impossible to form an open topped tray-like preform from cloth at 0/90 to the tray sides.

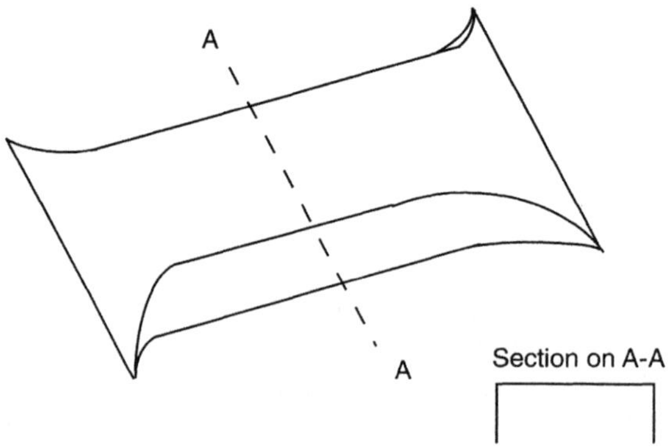

Figure 3.4 By folding on curved lines 3D shapes can be created with only highly localized in-plane deformations such that reinforcements of very low drape may be used.

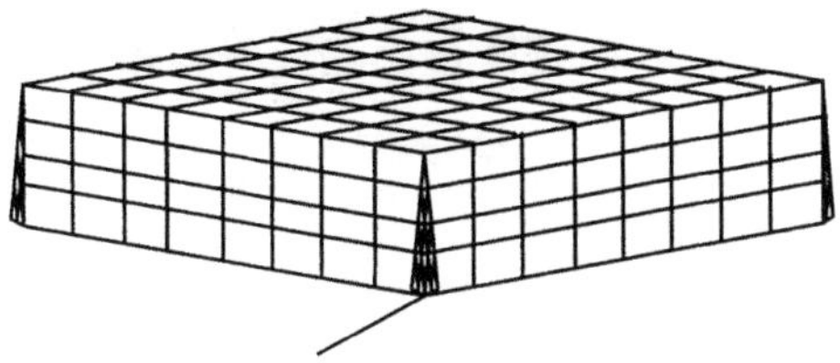

Figure 3.5 Attempting to form a tray from cloth at 0.90 to the tray edges will always lead to fibre wrinkling.

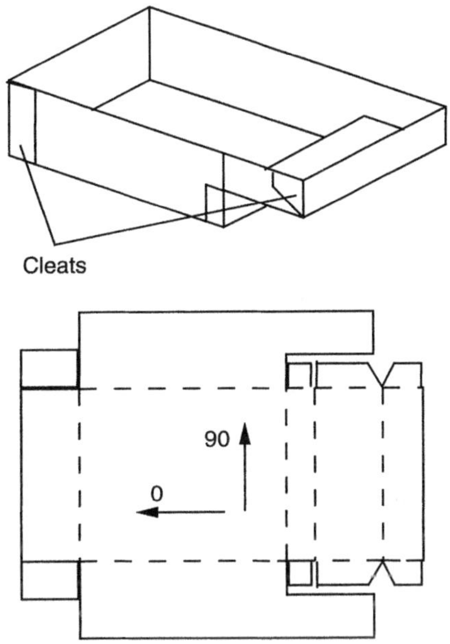

Figure 3.6 Cut and fold techniques can be used to make preforms in 0.90 materials, cleats at corners can be used to restore fibre continuity.

If the tray must be made at 0/90, cut and fold techniques must be used as shown in Figure 3.6. Figure 3.7 shows the sort of fibre angles that would be seen in the preform if cloth at +/–45 were to be used rather than at 0/90. In this case the preform would be drapeable up to some limiting tray height.

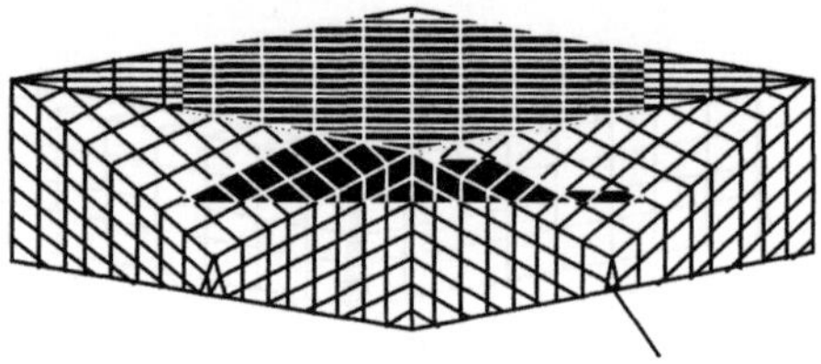

Figure 3.7 Wrinkles need not form if the cloth is at +/– 45 to the tray edge, at least up to some limiting height.

3.3 STEPS IN THE PREFORMING PROCESS FOR BOUND REINFORCEMENTS

The comments made below assume the use of thermally softened binders on balanced weave cloth in multi-ply stacks as the baseline process.

3.3.1 Preparation of stock sheets

Sheets of bound cloth are taken and laid up to the correct orientation and ply count. This stack is then heated and pressed to bind the layers together. In small-scale development work a flat iron may be used to bond the layers. At larger scales a platen press or heated vacuum table may be used, with the latter probably being the best option. The stock sheets can be stacked up from precut individual plies to produce developed shape preform blanks directly, or full width sheets can be made that must be cut down into preform blanks. The former may sometimes lead to improvements in reinforcement utilization, but the second approach requires less accuracy of layup and a lower labour input. In addition to constant thickness stock sheets it is possible to lay up stock sheets of variable thickness, these tend to be used more often as packers than as stock to be formed into double curvatures (Figure 3.8).

3.3.2 Preparation of preform blanks

The shape of the preform blank should be the flat developed shape of the preform, plus a preform trimming allowance. The developed shape is that shape generated when a well-formed preform is accurately trimmed to shape then heated and put back into a flat format. The determination of the flat developed shape is quite simple for biaxial cloth where deformation is essentially reversible up to large deformations. For other materials the flat developed shape could only be determined by trial and error methods. The preform trimming allowance is a function of the total

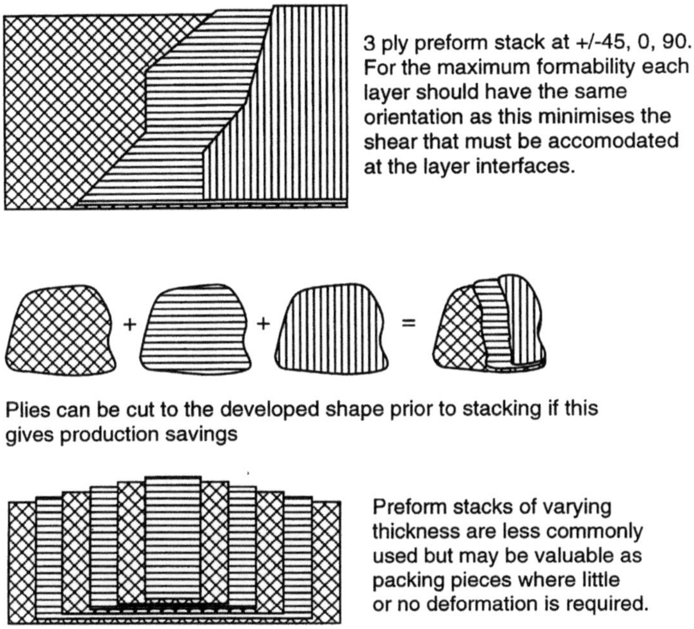

Figure 3.8 Various forms of preform stacks can be used depending on the requirements.

accuracy of the preform rig and the reproducibility of the forming process for the reinforcements to be used. Trimming allowances of the order of 1 cm seem to be about right for smaller preforms. The material wastage inherent in a trimming allowance can be minimized by net shape or near net shape preforming. To carry out net shape forming the forming process must be very reproducible, from the reinforcement deformation viewpoint and with regard to the alignment of forming tools. In order to maintain accurate alignment between forming tool and preform blank some tool location features such as extension tabs on the preform blank will be required. These tabs would generally be cut off before preform use, such that even net-shape preforms may require some trimming. Cutting of the preform blank shapes can be achieved in many ways. Simple hand cutting to templates would be the norm in development, with rule dies and clicker or roller presses being satisfactory in production. As with any reinforcement cutting, all the preforms of one stock type should be nested together for maximum materials usage. If holes are required in the final preform these can be cut out prior to forming if the holes are in undeformed areas of the preform, and such holes can then be used as tool alignment features. Any holes cut in areas of the

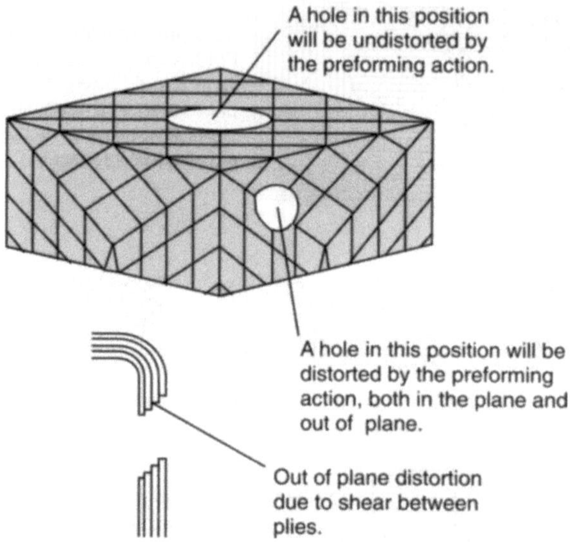

Figure 3.9 Holes should be cut through preforms prior to preforming only if no deformation will occur in the region of the hole.

preform that are deformed in the forming process may be distorted during forming and are best cut in the finished preform. See Figure 3.9.

3.3.3 Forming the preforms

The requirements are that the preform blank be heated to the binder's softening temperature, transferred to the press station, held in register with the forming tools and rapidly pressed to shape, being held in the preform tools until it has cooled sufficiently to retain its shape when released. The details of these processes depend on the design of the preform rig utilized and will be discussed later. Cycle times can be in the 1–2 minutes range. Generally the heating stage takes the bulk of this time, with transfer, pressing and cooling taking place in seconds.

3.3.4 Preform trimming

Preforms should be trimmed to size prior to storage and must, of course, be trimmed to size prior to use. The methods to be used depend to a large extent on the preform shape. If the trim line falls on a flat plane the preform can be cut using rule dies and these could be incorporated into the preform tool, assuming that the preform press generates enough force to cut the preform. For complex preforms hand trimming is commonplace, with or without some jigging assistance. The preforming tools can be designed so as to leave some indication of the trim line on the preform. Cutting dry fibre preforms inevitably gives rise to some airborne fibres and dust. Experiments with bright light beams passed over the cutting area have shown that airborne fibre levels are moderate if the trimming allowance is greater than the cloth's unit cell. If near net shape preforms are used where trimming allowances may be 1 or 2 mm a much larger amount of airborne fibre is liberated. In general it is recommended that trimming operations should be carried out in a suitable fume extracted area and this will be essential if trimming allowances are very small. Similar experiments at the ply cutting (by roller press) and preforming stages showed much lower levels of airborne fibre generation. These activities probably require no special precautions, beyond the normal high levels of ventilation required in composites production. For any specific plant it would be best to check airborne fibre generation from each activity prior to making decisions as to ventilation and extraction requirements, as small differences in processing may have larger effects on airborne fibre generation. One last point should be made about airborne fibres. If the fibres are carbon they are electrically conductive and can be drawn into electronic equipment by internal cooling fans. In my experience these fibres can lead to serious problems such as blown computer power supplies, making them a concern for equipment health as well as operator health and comfort.

3.3.5 Quality control of preforms

A fuller treatment of quality control in quality critical RTM is given in Chapter 13. QC checks on preforms would be expected to consider whether the correct press cycle is used, that preforms are fully bonded, lack defects such as wrinkles or tears and that dimensions are as required. As preforms are usually somewhat flexible, dimensional checking in a conventional way will be difficult. One approach is to take a splash from the preform tool, mark on this the trim line tolerances and to use this as a go/no go gauge.

3.3.6 Storage and handling of preforms

Preforms need to be stored in cool, dry conditions to prevent relaxation of the formed shape. They must be protected from contamination with dirt, dust, water etc. As with any form of composite reinforcement they should not be touched with bare hands (or even worse with hands covered in barrier cream). Even the most robust of preforms can be damaged by careless handling or storage and preforms should not really be stacked together for storage. One solution to preform storage is the adoption of kitting trays that hold all the preforms for a particular moulding. These are easily made by vacuum forming and should incorporate a recess for each preform to provide optimum support.

The use of such kitting trays permits the preform suite to be checked at a glance for completeness. Kit trays can also carry numbers to ensure that the mould loading sequence is self-evident. Preforms that are to be further built up into assemblies of preforms outside the mould tool, rather than loaded directly into the tool, should be treated in the same way.

3.4 PREFORMING EQUIPMENT

Preforming equipment performs several functions. It heats the blank, ensures that it is held in register with the preform tools and reacts the loads from the preforming action. There is a great deal of flexibility in how these functions can be achieved and in the overall shape of the equipment. Two major equipment types may be identified, shuttle rigs and carousel rigs.

3.4.1 Shuttle rigs

In these rigs either the blankholder or the heaters are shuttled back and forth (see Figure 3.10).[9]

In the first case the rig would have a fixed heater station and a fixed press station. The blankholder would be shuttled into the heater zone for

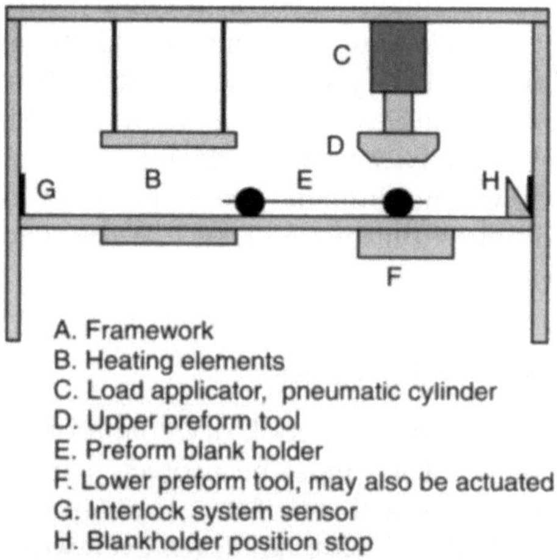

Figure 3.10 The main features of a shuttle type preformer are shown here, the level of interlocking and automation depends on required production rate.

the required time, then quickly transferred to the press zone. The gap between the toolfaces would only have to clear the blankholder, thus very little time is lost in tool movement. The heater station could be infra red, a conventional oven, hot plate or air circulation type. A fixed position heater zone gives the most flexibility of heater type. If the blankholder is fixed the heater(s) must be shuttled back and forth which tends to limit them to radiant heat types. Greater clearance between the blankholder and preform tooling is required which means more time lost in tool movement and leads to a requirement for longer stroke actuators. The advantages of a fixed blankholder are that the positional relationships between the tooling and blank are guaranteed; also, if the heaters are carried on hinged arms rather than sliding on rails, a more compact rig can be produced. The various shuttle actions can be carried out manually or pneumatic cylinders or other actuators can be used to give a higher level of automation.

When the blank has reached the correct temperature the shuttling is activated and the blank pressed. This is generally done between matched tools, but for some shapes single-sided tools used in conjunction with a rubber sheet and vacuum actuation can be used; so long as the draw into the tools is not constrained by the rubber sheet (Figure 3.11).

Tools can be actuated by pneumatic cylinders for speed and low-cost construction if, as is usually the case, forming loads are low. As the intent

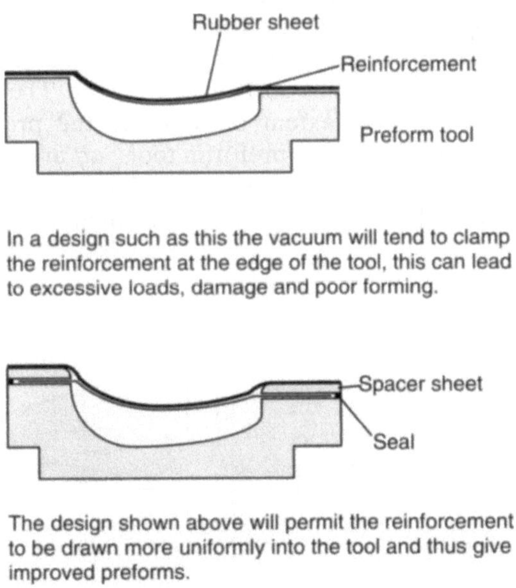

Figure 3.11 In some instances rubber-faced vacuum boxes can be used to form preforms.

is to operate on a rapid cycle time safety interlocking is required. Photoelectric cell safety cutouts may be preferable to conventional press guards as access to the working area is generally required for loading and unloading. If the blankholder shuttles, the whole of the press area can be isolated from human intervention by the use of an intermediate position for blank loading and preform unloading. Shuttle preform rigs are usually used above a preform size of a few hundred mm square and rigs with a working area up to a few square metres have been built. Cycle time depends on heating rate, which essentially depends on blank thickness and heat transfer rates. Cycle times of two minutes have been achieved, which compares very favourably to the length of time that can be taken to lay down several plies of prepreg on a complex form.

3.4.2 Carousel rigs

Carousel rigs operate with separate stations for heating, pressing and loading/extraction. Each of these stations are occupied at any one time, leading to a greater throughput rate. In the simplest case the three stations are arranged around a central shaft (see Figure 3.12).

The blank is loaded and rotated into the heater, a second blank is then loaded. After a preset time the carousel is rotated and the first blank is pressed while the second begins its heating cycle and a third is loaded. Cycle time depends on whichever of the three activities is the slowest. This is generally the heating, but if it is not then this stage must be carefully controlled to prevent overheating. Cycle times of one minute or less have been achieved with this sort of equipment. The guarding requirements depend to a large extent on whether the preforms have to be manually extracted from the preform tools or are carried off on the blankholder plates, which depends on the detailed design of tooling and blankholders as well as the shapes to be formed.

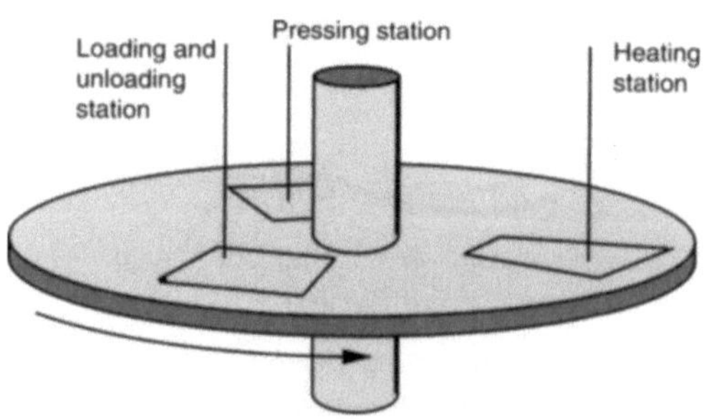

Figure 3.12 Schematic of a rotary preform machine. The rotating table must be provided with click stops to ensure correct alignment. Unlike the shuttle rig, each station can be filled at all times increasing the preform production rate.

Whichever form of preform rig is used care must be taken over the design of the attachment of preform tools to the rig, both in terms of alignment and tool changing. If a component contains, say, ten preforms the total process time may be 20 minutes plus tool setup time. In the context of advanced applications this is likely to be a small proportion of the total cycle time and thus one preform machine can service several mould tools, equally it is unlikely that production volumes would be sufficient to justify a separate rig for each preform type. Thus the tool changing time can greatly influence the optimum batch size for each type of preform, with regard to forming, trimming and kitting. It would be possible to develop automatic or semi-automatic tool changers, but it is unlikely that the capital costs of this could be justified, so the aim is to produce light tools that are easily and accurately interchangeable in the minimum time.

3.5 PREFORMING TOOLS

Preform tools are essentially light duty press tools. They can be made from a wide variety of materials; plaster, wood, tooling foam, casting resin, highly filled epoxy putty, rubber, GRP and metal have all been used. The choice of material depends largely on the number of preforms to be made and the complexity of the geometry. GRP has proven to be very good for production tools,[10] and various casting resins and moulding pastes have been widely used in development work.

A single preform is unlikely to represent the full thickness of the component to be made. The suite of preforms can be viewed as a set of offsets from the faces of the mould tool that nest together in 3D to form the component geometry (see Figure 3.13).

These offsets can be generated from CAD data and used to drive milling machinery to cut an initial set of preform tools from tooling foam. Alternatively, conventional mouldmaking techniques can be used to generate the tools from a master model. Figures 3.14 and 3.15 show this approach. From the master model the preform periphery is shuttered off and a shell cast in a suitable material. This shell is then built up with constant thickness wax sheet to the preform thickness required. The second tool half is then cast off the waxed shell and the two halves are mounted onto suitable frames for attachment to the preform press. The process is repeated for the other preforms. The suite of preform tools made in this way should not be made of too robust a material, unless either any reinforcement thickness changes due to preforming have been fully modelled or are trivial. When these initial preform tools are first used they will tend to print in regions of high deformation or great thickness changes. Some modification to the tool geometries is often required

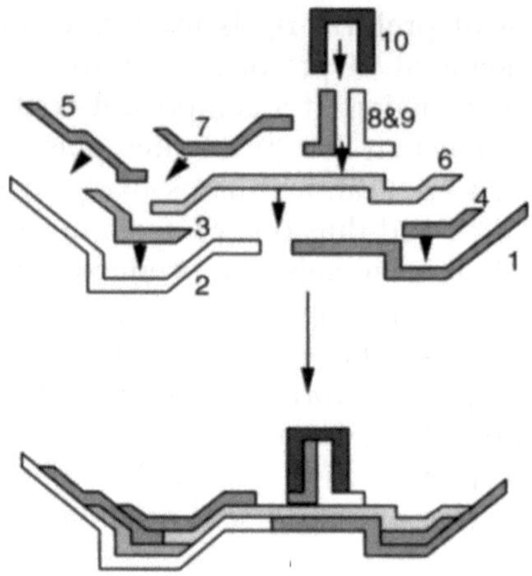

Figure 3.13 Exploded view of how preforms can be used to build up complex parts. (Preform thickness greatly exaggerated.)

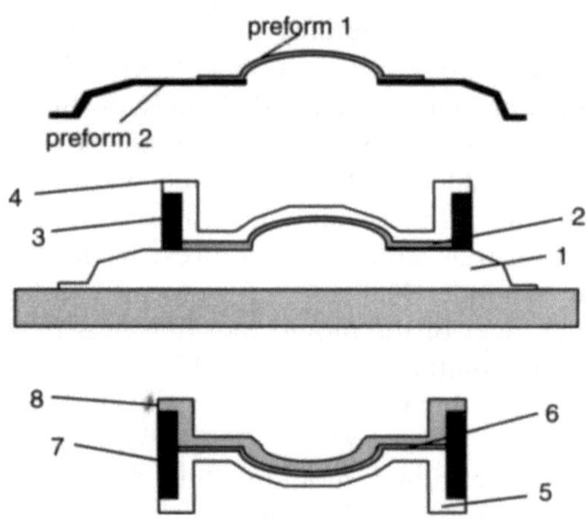

Figure 3.14 Stages in the manufacture of a set of preform tools (a)

1. Master model, in this case the model is the internal geometry.
2. Apply thickness wax to give external profile of preform 1, made oversize for a trimming allowance.
3. Shutter around periphery.
4. GRP 1st preform shell.
5. Invert 1st preform half.
6. Apply thickness wax to internal profile.
7. Shutter around periphery.
8. GRP 2nd preform shell.
9. (Not shown). Mount preform shells on boards.

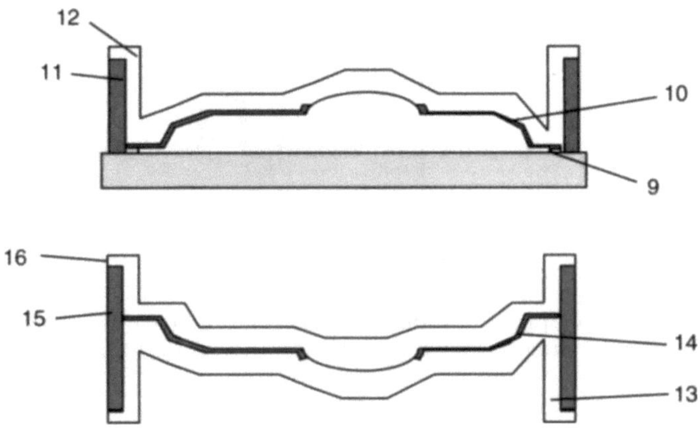

Figure 3.15 Stages in the manufacture of a set of preform tools (b)

9. Extend periphery of model to provide trimming allowance.
10. Apply thickness wax to model over running the required trim position at both edges.
11. Shutter around the periphery.
12. GRP 1st preform half.
13. Invert 1st preform shell.
14. Apply thickness wax.
15. Shutter around periphery.
16. GRP 2nd preform shell.
17. (Not shown.) Mount preform shells on boards.

If CAD data is available preform tools can be milled directly from CAD as offsets from the tool geometry, ensuring that a trimming allowance is made.

to cope with these changes in dimensions and this is easier to achieve on softer materials. When the accuracy of individual preforming tools and the fit of the total preform suite has been established production preform tools can be procured. GRP, metal and high performance moulding compounds such as SiC or titanium filled epoxies have proven to be acceptable. As changes in preform details may be necessary during the preform development process it is preferable to generate final preform and component production drawings after the completion of the preform development process.

Not all geometries can be manufactured in simple preforming tools without damage to the reinforcement. In this case the preform may be split into two to permit forming or sequential forming or double action preform tools can be used. This is similar to the case of sheet metal stamping of complex shapes, where several actions may take place within a single press tool.[11]

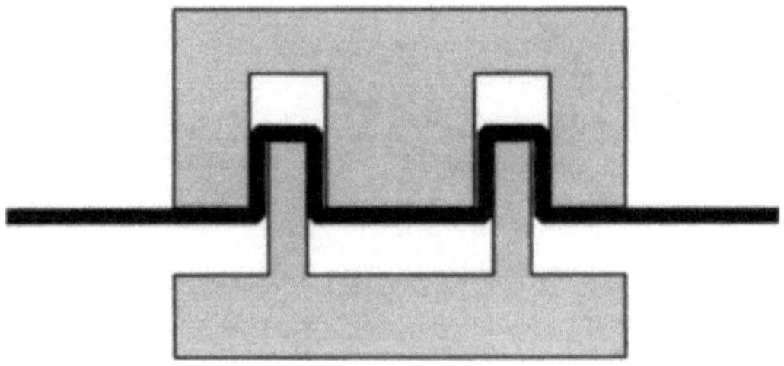

Figure 3.16 Simple preform tools can sometimes damage the reinforcement if it must be drawn a great distance through the tool as it closes.

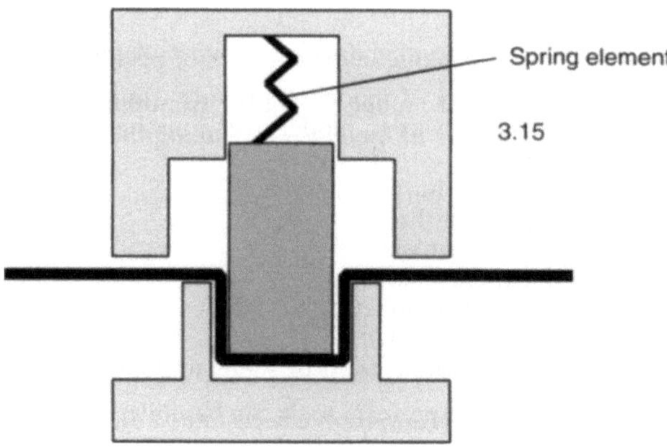

Figure 3.17 Double-acting preform tools can be used to avoid damaging the reinforcement.

Figure 3.16 shows how the reinforcement may be damaged by high forces generated by having to draw the material through a simple tool. The preform shown could be made in two sequential steps, either in tooling or by off-line preshaping of the reinforcement, but reheating the preform would be difficult (at least with radiant heating).

Figure 3.17 shows a double-acting tool that can be used to form the required shape in a single action. The centre section closes first to form that section of the preform, as the tool closes further the outer sections perform their forming actions to generate the preform without excessive forces. In principle multiple action tools could be used with many more than two actions but it is likely that the added complexity would soon become unmanageable. When discussing reinforcement deformability in section 3.2 it was noted that all reinforcements require a characteristic number of free edges to deform without defects being formed. In double-action tools the first action will restrict the number of free edges available and the deformation limit at which wrinkling or folding occurs may be lower in the second action than in the first action. In some cases this may make it unavoidable to split the preform to achieve adequate quality.

REFERENCES

1. Pelstring, R. M. and Madan, R. C. (1989) Stitching to improve damage tolerance of composites, *34th International SAMPE Symposium*, May, 1519–28.
2. Potter, K. D. (1979) The influence of accurate stretch data for reinforcements on the production of complex structural mouldings. Part 1. Deformation of aligned sheets and fabrics, *Composites*, July, 161–7.
3. Potter, K. D. (1979) The influence of accurate stretch data for reinforcements on the production of complex structural mouldings. Part 2. Deformation of random mats, *Composites*, July, 168–73.
4. Okine, D., Edison, R. and Little, N. (1987) Properties and formability of a novel advanced thermoplastic composite sheet product, *SAMPE Engineering Series* **32**, 1413.
5. Long, A., Rudd, C., Blagdon, M. and Smith, P. (1996) Characterising the processing and performance of aligned reinforcement during preform manufacture, *Composites*, Part A, 27A, 4.
6. PDA Engineering International Ltd (1994) *P3/Laminate Modeller Datasheet,* Basingstoke: PDA Engineering International Ltd., Basingstoke, Hampshire, RG21 2JX, UK.
7. Potter, K. D. and Robertson, F. C. (1987) Bismaleimide formulations for resin transfer moulding, *32nd International SAMPE Symposium*, April.
8. Potter, K. (1983) High rate reforming of thermoplastic laminate into small, complex components; a process study, *Proc. 4th International Conference*, SAMPE European Chapter, SAMPE, 103–12.
9. Horn, S., Buckley, D. and Seroogy, K. (1990) High volume, highly automated preform process for RTM and SRIM, *45th Annual Conf. Composites Institute*, SPI, Feb., Session 9-C.
10. Jones, W. and Johnson, J. (1980) A resin injection technique for the fabrication of aero-engine composite components, *Proc. Symposium: Fabrication Techniques for Advanced Reinforced Plastics*, April, Salford: IPC Science and Technology, 40–7.
11. DeGarmo, E. P. (1969) *Materials and Processes in Manufacturing*, 3rd edition, London: Macmillan.

4 RTM mould tool design

4.1 INTRODUCTION

It is almost impossible to overestimate the importance of properly designed and manufactured tooling in composites processing. Even in the case of relatively simple, single-sided tools such as are used for autoclave moulding I have seen many instances where inadequate tooling was one of the greatest sources of scrap or reworked components in a production setting. In most variants of RTM the tooling is solely responsible for the geometry and tolerances of the part. The tool dictates how the resin will enter and leave the part and in very large part controls the filling process and thus the quality of the components. The tool must also react all the loads from mould closure and resin pressure and provide the heat required to cure the resin. The longevity and costs of the tooling make a major contribution to production economics. Its weight and ease of handling – with regard to preform loading, clamping, resin porting control and resistance to resin pressure, part ejection and transport – are major factors in the design of the production engineering approaches to be used.

For all these reasons it can be expected that the design of adequate RTM tools will be a more complex task than tool design for autoclave or other single-sided tools. Mould design and component design are inextricably linked to each other and to production engineering. Tooling approaches must be designed and planned with as much care as that given to the components themselves. Equally, component and tool design are heavily interlinked. For example changes in assumed component Vf% levels can have dramatic effects on mould closure forces which might affect materials of construction and clamping arrangements and thus cascade through the whole of the production engineering and ultimately costing assumptions. Because of this sensitivity the tool design and component design processes must proceed in parallel, rather than the component being fully designed before tooling is considered. Implications of tooling design decisions on the potential production engineering approaches should also be considered as they arise and not be left to the end of the design process.

The baseline process which will be considered here is that of fixed cavity RTM in which the part geometry is defined and controlled by essentially rigid tooling elements. Other variants are possible such as the use of flexible bags for one face of the tool, or inflatable internal mandrels etc. These options can sometimes assist in overcoming some of the problem areas in the design of RTM tools, but at the cost of losing some control over part thickness and limiting the complexity of the available part geometry. If the principles for the design and construction of rigid tools are known, these can be transferred across to the other tooling types, for this reason the emphasis here is on rigid tooling.

The author's experience with the design and operation of RTM tools has covered a wide range of materials, types of tool, clamping, sealing, heating, injection and ejection options; in both development and production environments. This experience has mostly been gained in components where the typical requirements are for complex shapes, tight dimensional tolerances, high-fibre contents and high-quality requirements. This experience has undoubtedly coloured my responses to different tooling approaches and materials. There may therefore be statements made that others disagree with due to their different experience. In many cases these disagreements may be based on different frames of reference, for example, a tooling method which is considered to give good tolerances in one market sector might be wholly inadequate for another. In other cases my ignorance of other market sectors may have led me astray, for which apologies are offered in advance. Equally, RTM is still a rapidly expanding field, with new developments coming at a rapid pace and thus new developments may not always be covered here. Lastly, the challenges and problems of any particular component might simply not have arisen before. Unfortunately, RTM has not been in use for long enough that standard solutions have been developed to all potential problems. Tool design is very much a real design task, rather than the application of standards. It is to be hoped that sufficient information is provided here that novice designers of RTM tools can respond to the problems with which they are presented and progress beyond the current state of the art.

4.2 TOOLING MATERIALS

One of the first questions to be answered is: what will the tool be made of? To answer this question raises many issues of the material's chemical, mechanical, physical and manufacturing properties. The sort of properties that might influence the design are noted below.

Three general classes of approach to making an RTM toolface might be defined. The direct method is to machine the mould cavity directly into blocks of material, using CAD/CAM, copy milling from a master or other methods. Materials that might be used are blocks of polymers, tooling foams, metals and machinable ceramics. The use of stereo-lithography techniques to produce prototype tooling would also be classed as a direct method.

Indirect methods utilize a master model, onto which the tooling faces can be cast or laminated. Materials that might be used include: castable polymers, including filled forms; GRP and CFRP; castable ceramics and concrete; nickel electroforms or sprayed metal etc. In some cases the actual toolface might be a gel coat rather than the main tooling material, adding extra complexity to the considerations. In other cases a light toolface may be backed up by casting a mass of a different material, again adding complexity.

Lastly, hybrid methods might include near-net casting of materials such as cast iron or aluminium, followed by NC machining of the actual toolface. The support of a lightweight toolface by metallic, or other, framing materials to support and stiffen it might also fall into this category.

Whatever the materials and methods used, some data are required in many areas, in order to define how to meet the total use requirements as cost-effectively and securely as possible.

Chemical properties include; compatibility with resins and their curing agents and any solvents and release agents to be used, plus any tendency to corrosion in the production environment.

Mechanical properties include: strength, stiffness and resistance to handling induced damage. This latter is not a simple matter of toughness expressed as fracture energy etc. For example, aluminium is tougher than cast iron but can be much easier to damage by poor design or handling practices. Equally, for very critical components, there might be a preference for a toolface that is either perfect or obviously damaged. In this case a ceramic might be better than, say, GRP, where delamination or similar damage might affect quality or dimensions without being immediately apparent.

Physical properties of importance might include: density; coefficient of thermal expansion; thermal conductivity; specific heat; porosity; maximum use temperature etc.

Manufacturing properties might include: ease of machining; achievable tolerances; surface finish; ease of integration with presses, ejectors and injectors; acquisition time; cure or thermal shrinkages or other dimensional changes; maximum size; minimum and maximum gauge limitations; availability of suitable suppliers; their workload and competitive positions etc.

In addition there are more complex properties that build up from a consideration of the simpler properties and may also be very much influenced by other elements of the total production engineering. These include: longevity, maintainability, repairability, ease of modification, limitations on geometry or complexity etc. For example a toolface that had adequate longevity in a low-temperature curing product line making objects of simple geometry, where mould loading and unloading were automated, might be totally unsuitable in a manually operated line making complex products at high temperature.

To conclude, there are three general approaches to tool-making, a very wide range of material types (and a much wider range of specific grades) and more than 20 different properties that might be of critical importance. It would be ideal if a simple decision tree could be drawn that led designers straight to the best choice of materials or design type; unfortunately this seems not to be a realistic option at this time as the possibilities are so numerous.

Nowhere in the above listings has the cost of tool acquisition been mentioned. This is because meaningful figures are almost impossible to define, especially if cost per moulding is the criterion, as production volume is then critical. Even in the absence of a requirement to define cost per moulding the available equipment and technology can dominate the issue. If a full set of CAD data and an NC mill are available then direct machining of the toolfaces from metal is often cheaper and quicker than the use of indirect methods. If a master model of the component is available the reverse is quite likely to be true. In addition, the acquisition cost of the tools is only one element of the total costs. For example, a decision to use heavy metallic tools with temperature control by hot oil circulation has profound implications on the costs of peripheral equipment, tool handling and manipulation. Once these costs have been met for one product it may be more cost-ffective to use similar technology in future tools as they can share facilities, even if each individual tool is more costly. The costs of parts inspection, finishing operations and even scrap and rework rates can vary with tooling type (due to tolerance limitations, tooling degradation etc.) so that the only real indicator of tool cost is the total production cost. This is not to say that tooling acquisition cost is unimportant, but rather that before acquiring apparently cheap tools a great deal of thought is required as to any costs that may appear elsewhere as a result of the decision.

Despite all the above it is necessary to attempt an indication of the relative merits of materials and approaches. As noted earlier a tooling material and approach that gives adequate performance in one area might be wholly inadequate in another. The key, as always, is fitness for purpose and materials cannot realistically be ranked in terms of absolute properties to give a 'best' material. Unfortunately it is very difficult to remove all trace of one's own prejudices and experiences from such judgements and my past experiences will no doubt be visible in the listings given below. Each material will be considered in turn and the main features, manufacturing options and positive and negative features will be noted. For materials that can be handled by several routes the differences between the options will be noted.

4.2.1 Filled and unfilled polymers

These include precast blocks or other shapes of polymers such as PTFE, nylons and other chemistries. Mass cast systems such as filled epoxies, polyesters, polyurethanes, syntactic foams or methacrylic polymers also fall into this group. Some mass cast systems specifically manufactured for composites tooling are available.[1]

Specific gravity: From about 0.6 for syntactic foams to above 2 for systems that are highly filled with dense fillers.

Strength: Up to about 100 MPa, very dependent on type of polymer and usable strength may be lower due to yielding or creep effects, especially towards the use temperature limits.

Modulus: From about 0.5 GPa for PTFE to around 5 GPa for highly filled systems.

Toughness: In terms of minor impact damage polymers can exhibit high toughness, edges on tools made from mass casting systems may be prone to chipping. Thin edges machined on toolfaces will be particularly susceptible to damage. Except for such systems as highly filled epoxies with hard fillers or some aluminium filled mass cast systems the toolfaces are very easily scratched or marked by rough use. Great care is needed in cleaning to avoid such damage. It is good practice to ensure that any tools such as scrapers used to clean tools are softer than the toolface.

Tolerances: For machined tools that have been well-stress-relieved, tolerances can be quite good. All polymeric systems tend to have relatively high thermal expansion coefficients that can lead to a loss of tolerances if the tool temperature is not well controlled. Cast toolfaces cannot be better toleranced than the master model and shrinkages on cure should be taken into account if good tolerances are required.

Surface finish: Surface finish is never likely to be as good as for a metal tool as the materials can be hard to polish and harder still to keep free of scratches and other minor damage. The harder, filled, materials would be expected to be better than unfilled materials in the long run.

Longevity: This is a complex factor, depending on tolerances and surface finish etc. The lower the component requirements the longer the tool would be expected to be usable. In general such tools are more likely to be used for prototyping and short-run production than for large series production, although some mass cast variants are claimed to be suitable for long production runs.

Tooling details: Some sort of support framework would often be required unless tools were very small. Seals would be expected to be gaskets rather than 'O' rings for cast tools. Metal inserts could be used for injection points or ejectors so long as the tool was not used at elevated temperatures where the differences in expansion coefficient would be a problem. Problems might be experienced with bonding in metal inserts to machined toolfaces.

Coefficient of thermal expansion: In the range $30–100 \times 10^{-6}/°C$. This is much higher than that of any composite material and limits the use of such materials to low-temperature cure; or at least to constant temperature tools. In some cases high expansion coefficients can be of advantage in high-temperature curing systems, e.g. when used as cores in parallel-sided bores, where the high expansion coefficient makes demoulding easy. PTFE has often been used for this sort of duty.

Thermal conductivity: In general very low, from about 0.1 to 0.5% of that of aluminium. This can cause problems both with the build-up of exothermic heat in series production and with heating the tool to cure the resin above room temperature. Heating fluid pipes can be cast into mass cast systems. In order to achieve temperature uniformity a close pitch of heaters and/or a substantial thickness of material above the tubes would be required. Very low conductivity systems need great care in heating, if electrical heaters are used localized hot spots can severely degrade the tool's lifetime.

Specific heat: About 2000 J/kg/°C

Maximum use temperature: This would depend on many factors such as tolerances, heating requirements, mechanical loads etc., it would be unlikely that such systems would be used much above 100°C, although plastic elements might be used in other tools at a higher temperature.

Gauge limitations: Machined plastics can be very thin, gauge would depend on distortions, handling, deflection limits etc. For cast systems minimum gauge seems to be about 15 mm for reasons of manufacturing and ease of handling.

Acquisition time: Fairly small mass-cast tools might be built from a master model in a few days, perhaps one day for each face and another day for fitting out and framing. Most plastics can be readily and quickly machined if this route is being followed.

Repair: Chips or similar damage in mass-cast tools could be repaired with filling compounds or the original mass-casting materials. Machined plastic toolfaces based on thermoplastic chemistry are likely to be impossible to repair.

Modification: Beyond small changes, such as easing off high Vf% areas, it would be difficult to make major modifications.

Chemical compatibility: Some types such as PTFE need no release agents and have no compatibility problems with solvents or resins. Curing systems will require release agents and some may be sensitive to solvents and/or the styrene used in polyester resins, although they would be unlikely to be attacked by the acid catalysts used in phenolics. It would be prudent to ensure that the compatibility of tooling materials and any solvents is checked prior to the selection of this sort of tooling.

Size limitations: Machined tools are likely to be small. Mass-cast tools could be larger, depending on exothermic heat generation in manufacture. The dimension limited by exothermic heat is essentially the thickness and not the plan area. Therefore, if the toolface is supported in such a way that it does not carry major loads and can be treated as a thin shell, tools of a large plan area could be built; although handling large quantities of mass cast materials may be difficult.

Limits to complexity: These are set as much by handling, tolerances both for the component and for the tooling elements in multipart tools, toughness etc. as they are by the physical ability to generate the geometry. Tools of this nature are unlikely to be of the highest level of complexity.

In general, plastic toolfaces would be expected to be used mostly for relatively small items of low to moderate complexity, and where mould closure and injection loads and use temperatures are low to moderate. As noted above, mass-cast materials could be used for larger tools, so long as the loads are carried outside the toolface. Transparent plastic tools are also sometimes used to study the flow of resin through the tool, as an aid in prototyping and the design of production tools. In this case machining from perspex or the use of embedding resins are the normal production routes. In addition PTFE is sometimes used as a tool element to generate bores inside a component where its high CTE, low modulus and lack of adhesion to resins can be of great benefit.

4.2.2 Tooling foams

These materials are light and stable materials intended for the manufacture of master models. The properties that make them useful in these roles can also occasionally make them suitable for use in tooling.

Specific gravity: Around 0.4 to 0.6.
Strength: Around 20 MPa in compression, easily dented.
Modulus: Around 1 GPa.
Toughness: Easily cut, dented or marked.

Tolerances: These materials are designed to be stable, therefore machined tolerances will be good, degradation might occur over time with loading.
Surface finish: A very good surface finish can be achieved, but this would not be robust in service.
Longevity: Expected to be low, due to the ease of damaging the material.
Tooling details: Unless tools were very small good support would be required, mould furniture could easily be bonded in. Seals would be more likely to be gaskets than 'O' rings as the forces to seal these would be likely to dent the foam.
Coefficient of thermal expansion: As for plastics.
Specific heat: As for plastics.
Thermal conductivity: Taking into account that the foam must be thick to provide adequate stiffness it can be assumed to be a near-perfect insulator. For this reason it is reasonable to assume use only at or close to room temperature.
Maximum use temperature: In principle up to around 100°C, in practice room temperature (see above).
Gauge limitations: The material has low strength and stiffness, therefore thicknesses below around 30 mm are unlikely.
Acquisition time: Tooling foam is designed to be quick and easy to work by mechanical or manual means. If CAD data is to be used to machine toolfaces it is a common practice to cut a version quickly in tooling foam to check that the programming is correct. These try-out toolfaces can then be used for initial prototyping trials to check out mould loading and other details prior to cutting hard tools. In this case the softness of the foam can be an advantage as any high Vf% areas will 'print' on the foam.
Repair: Tooling foam is easy to repair with filler pastes.
Modifications: Minor modifications are easy to achieve by adding or removing material. Major modifications can be achieved by splicing in a section of foam and remachining.
Chemical compatibility: The surface of tooling foam is porous and would have to be sealed with some form of varnish prior to use, the compatibility issue is more with the material chosen for this duty than with the foam itself.
Size limitations: Tooling foam is available in large sheets that could be spliced together to make very large blocks. The limitation on size is therefore associated with the available machining equipment rather than the material.
Limits to complexity: The limiting factor would be the fragility of thin sections of foam. If NC machining is used the availability of good tolerances would allow multipart tools to be made.

In general tooling foam is most likely to be used for very short run and prototyping activities and for tool tryout when NC machining is the preferred manufacturing route for production tools. Tooling foam is also the material of choice for master models for indirect methods of manufacturing tools.

4.2.3 GRP

Unsurprisingly, GRP is very commonly used for the manufacture of RTM tools, especially in the general products sector where glass is the dominant reinforcement fibre. Hand lamination onto a gel coat and a master model is the norm, although prepreg may be used. Some suppliers[2] sell a complete package of all the materials required, plus ancillaries and mould furniture for this type of construction; training courses are also available. Some toolmakers use core materials in the layup to stiffen the tool and minimize print-through from supporting steelwork. In this sort of tool electric heater mats or heater pipes may be laminated in to control the tool temperature.

Specific gravity: 1.5 to 2 depending on glass content.

Strength: From about 150 to 400 MPa depending on layup and glass content.

Modulus: From about 7 to 20 GPa depending on layup and glass content.

Toughness: GRP tools can be quite robust, but gel coats are easily damaged by scratches and impacts can cause delaminations or loss of vacuum integrity. Moulds that see a lot of thermal cycling can also slowly degrade. Overheating or excessive pressurization can also lead to damage and reductions in tool life. Fine details on the toolface tend to be largely gel coat and small external radii (such as might be seen on multipart tools) are easily damaged and hard to manufacture.

Tolerances: In principle identical to those of the master model. In practice the small geometrical changes associated with cure shrinkages and thermally induced shrinkages tend to degrade the tolerances somewhat. The best achievable tolerances are below those achievable with machined tools, which can cause fit problems if a suite of parts is being made.

Surface finish: As made the gel cost replicates the master and can be improved by polishing. With prolonged use the surface finish tends to degrade.

Longevity: Longevity is a function of the criteria chosen for acceptable components and, as such, hard and fast figures are impossible. Tool lives of thousands of cycles have been claimed. The author has also seen complete destruction of a GRP tool in one cycle, due to excessive tool heating, as well as cases where maintenance and repair became uneconomic in less than 100 cycles. This is most likely when high pressures or temperatures are used.

Tooling details: GRP tools generally require support frames, usually steel. For tools used at low temperatures steel inserts can be moulded in for injectors, ejectors, clamps etc. If tools are cured or used at high temperatures mechanical attachment of such metalwork may be preferred to avoid thermal mismatch strains and possible disbonding.

Coefficient of thermal expansion: In plane expansion is likely to be in the range $15\text{–}20 \times 10^{-6}$/°C, with through thickness expansion around twice as high. This difference between in and out of plane expansion drives the geometry changes that can occur when GRP is thermally cycled.

Thermal conductivity: In the region of 0.5% of aluminium.

Specific heat: In the region of 1000 J/kg/°C.

Maximum use temperature: This depends on the resin type used. Basic polyesters or low temperature cure epoxies are unlikely to offer adequate performance much above 60–80 °C. High-performance resins could be used up to about 150 °C, generally these would be prepreg systems and CFRP would be preferred to GRP. To achieve high-temperature performance generally requires high-temperature cure; some epoxies are now available that will precure at low temperatures and can then be slowly postcured off the master model to increase the use temperature without risk of damage to the master model.

Gauge limitations: It is unlikely that much less than 10 mm of GRP will give adequate mechanical performance in a tool. This figure would be reasonable for small tools or those in which the geometry stabilizes the structure (e.g. hemispheres), if the toolface is loaded in bending by the injection pressure some support would be needed for such thin toolfaces.

Acquisition time: Accurate layup of GRP tools is a time-consuming business, but corners cannot be cut if good quality is to be achieved. Postcure times can also be quite prolonged. While tools can be made quickly the highest quality tools might take a few weeks of elapsed time to produce.

Repair: Minor damages to the gel coat can be repaired with a suitable tooling paste; major damages such as delaminations are very difficult or impossible to repair. Widespread crazing or surface damage to the gelcoat would generally result in scrapping the toolface. In this case the supporting steelwork would normally be re-used.

Modification: Only the most minor modifications to GRP tools are generally possible without seriously affecting the mould's performance or reducing its longevity.

Chemical compatibility: If resin and gel systems that are accepted for tooling use are specified, problems in this area would not be expected. GRP toolfaces are often used with acid cure phenolic systems that would attack most metals.

Size limitations: In principle GRP moulds can be made by hand layup to any size required, moulds with a plan area of several square metres have certainly been made and used with good results. As tools become larger the requirements for support steelwork become greater. Problems have been reported with excessive deflections in some large GRP tools that have had inadequate stiffening.

Limits to complexity: Very fine details that are mostly gel coat will not be robust in handling, multipart tools can be made, but tool designers must be aware that tolerance buildup can lead to a lock-on situation, draft angles may need to be increased in complex tools to allow for this.

GRP tools are very widely used in the production of GRP mouldings by RTM, where production volumes are low to moderate (e.g. 100s to low 1000s). Most of these uses are not for highly stressed or high-volume fraction mouldings and the limitations of GRP tools in terms of tolerances etc. are not then a problem in most cases. The main advantage of hand-laminated GRP tools in this general moulding sector is that the skills required to make the tools are already available within the workforce, so that remanufacture of damaged or degraded mouldfaces is seldom a problem. There are also many specialist manufacturers of GRP tools.

4.2.4 CFRP

In contrast to GRP tools, CFRP tools are generally made from prepreg via autoclave or vacuum bag routes rather than by hand-lamination methods, and gel coats are not used. They thus tend to find more favour with organizations possessing skills in prepreg moulding, although specialist manufacturers of composite moulds can usually handle prepreg as well as hand lamination.

Specific gravity: 1.5
Strength: Up to 500 MPa
Modulus: About 35– 50GPa
Toughness: Surface durability of CFRP laminates is rather better than that of gel coats or foam or plastic tools. Impact toughness of CFRP tools is poor, tools can easily be damaged or delaminated by dropped tools or similar low energy impacts. Great care should be taken to avoid such impacts in service. CFRP tools are rather more likely than GRP tools to be used with high levels of evacuation, vacuum tightness is also likely to be rapidly degraded by impacts and sometimes by

prolonged thermal cycling. Recent developments of more impact resistant tooling prepreg give improvements in this area.[3]

Tolerances: As before, tolerances reflect those of the master model. CFRP prepreg tools seem to suffer rather less from distortion than hand-laminated GRP tools, but the use of as low a cure temperature as possible is recommended to avoid thermally activated shape changes, and some tooling prepreg systems have been formulated for reduced cure temperature.[4]

Surface finish: Surface finish can be excellent and is rather less susceptible to degradation than gel coat surfaces.

Longevity: Rather better than for GRP tools (when both are used with high Vf% and elevated temperature cures), largely due to improved mechanical performance and the use of improved resin systems. Loss of vacuum retention due to thermally induced cracking within the laminate can be a problem, a system is now available that claims to have solved this problem to at least a thousand high temperature use cycles.[3]

Tooling details: CFRP tools are less likely to be used with steel framework than GRP tools (if framing is required CFRP sheet is often used), problems of thermal mismatch if metal items are laminated or bonded into the structure are more likely than with GRP.

Coefficient of thermal expansion: About 4×10^{-6} in plane, essentially perfectly matched to multi-axial CFRP mouldings, and very badly matched to anything else.

Specific heat: In the region of 1000 J/kg/°C.

Thermal conductivity: Around 0.5% of that of aluminium.

Maximum temperature: CFRP tools can be used up to 180 °C with epoxy based tooling prepregs, higher temperatures might be available with BMI prepregs but longevity would be expected to decline. CFRP tools in aerospace are often oven heated to avoid having to incorporate heating systems. This is acceptable if long cycle times do not cause scheduling problems. Outside aerospace it would be more likely to incorporate some means of heating the tool.

Gauge limitations: Typical toolface thicknesses might be about 6–10 mm. If thicker material was required for loading reasons then framing or cored panels would be used.

Acquisition time: It would be expected that the acquisition time for CFRP prepreg tool would be a little longer than for GRP.

Repair: Only very minor surface blemishes are easily repaired with tooling paste, impact damage or similar damages might be temporarily repaired by injection of resin but the repair would not be expected to be lasting and replacement of the toolface would be required.

Modification: As for GRP tools it is unlikely that any major modifications to a CFRP tool could be made reliably.

Chemical compatibility: The fully cured epoxy resin systems used for CFRP tools would be expected to be very resistant to any solvent likely to be used in a production environment, and no problems would be expected with any resin type.

Size limitations: As CFRP tools generally require heat curing the size limitation is more or less set by the availability of suitable heating facilities.

Limits to complexity: Tight radii are very difficult to make reliably in CFRP toolfaces, multipart tools have been used successfully but, as for GRP, tolerance build-up can be a problem. One of the reasons for using RTM as a production technique is that it can provide better quality in complex geometries than other processes such as autoclave moulding. Thus CFRP is not likely to be used for the most complex RTM tools as this would lead to either limiting the complexity of the tool to that achievable with autoclave moulding or accepting a lower moulded quality in the tool than the component. This may not always be the case, but the most complex tools tend to be metal.

CFRP tooling is now very commonly used in the manufacture of CFRP components via autoclave and vacuum bag moulding and has also been used in RTM processing. Its use is likely to be limited by the questions of complexity raised above.

4.2.5 Nickel electroforms

In this process a master form is taken and an electroform (a very thick electroplating) is made from the master. This can then be used to make a reverse master for the other mould half and the process is repeated. The prime advantage is that a hard metallic surface can be taken from a master model without the requirement for machining. The density, strength etc. of electrodeposited nickel may vary from that of normally processed nickel, the values quoted here are for normally processed nickel.

Specific gravity: 8.9
Strength: 300 MPa
Modulus: 200 GPa

Toughness: The metal surface will be very much harder than the materials considered earlier, and much more resistant to minor damages such as dropped tools. Larger dropped objects can still cause severe damage by denting the surface.

Tolerances: As for any indirect process the best tolerance that can be achieved is that of the master model used. Special 'bath masters' have to be used for electrodeposition, if these are taken from a conventional master this represents another opportunity for tolerance degradation. On large tools it is unlikely that 'O' ring sealing tolerances can be achieved and gasket seals will be used. Support frames often include some provision for adjustment of the mould closure face.

Surface finish: The surface can be polished to an excellent level.

Longevity: With care and avoiding damage to, or overpressurization of, the toolfaces the tools should have a very long working life.

Tooling details: Inserts etc. can be introduced at or after electrodeposition. The electrodeposited shells themselves may not be adequately stiff and supporting frames of steel can be used; alternatively the faces could be backed up with cast aluminium, GRP or other similar materials.

Coefficient of thermal expansion: 13×10^{-6}/°C, midway between the in-plane expansion coefficients of GRP and CFRP and generally acceptable for use with either.

Thermal conductivity: About 25% of that of aluminium.

Specific heat: 460 J/kg/°C.

Maximum temperature: In excess of any of the currently used resin systems for RTM. Any backing material used may have a lower maximum use temperature.

Gauge limitations: The shell thickness is directly related to the bath residence time; the usual shell thickness will be in the 4–6 mm range.

Acquisition time: The first stages are similar to those for cast or GRP tools, the electrodeposition time will be substantially longer than the casting or layup time so that the total tool acquisition time might be 10 weeks or more. Only a limited range of vendors exists for electrodeposited nickel tools; most of these tools are used in autoclave moulding and it is not known how many vendors have the required experience with RTM.

Repair: Electrodeposited nickel tools would be expected to require less repair than many other tool types, it is difficult to see how any repairs other than very minor ones could be achieved in practice.

Modification: In principle very small amounts of material could be machined away for modifications; more major modifications would be expected to be difficult to achieve, due to the thin shells.

Chemical compatibility: No problems would be expected with resins or solvents, the toolface would require a wax sealant prior to use to fill any residual surface porosity from its manufacture.

Size limitations: While tools could in principle be sectionalized, difficulties might be expected in maintaining smooth transitions and vacuum integrity. For practical purposes size might be assumed to be limited by the available bath size, which will change from vendor to vendor.

Limits to complexity: The electroplating process, by its nature, produces shells rather than solid objects; multipart tools (for example to relieve undercuts) would therefore have to contain non-electroform elements.

Overall nickel electroform tooling might be considered for moderate areas and not too complex mouldings where a lightweight toolface of good long-term performance was required. Because of the relatively good thermal conductivity, ovens might be used to heat the tools for small-production volumes, for larger volume production heater fluid circulation in pipes bonded onto the external surface might be an easier option than attempting to fix electric heating elements to a thin shell. Nickel electroform tools of a plan area of up to about 10 square metres have been used in automotive RTM production, e.g. by Concargo Ltd of Weston Super Mare, UK.

4.2.6 Ceramics

This group is taken as including machinable and castable ceramics (including graphite), plaster and concrete. Such a wide grouping, from the cheapest to the most expensive tooling materials, does make it more difficult to be specific but most of these materials share some common properties.

Specific gravity: About 2.3 to 3.3.

Strength: Machinable ceramics and graphite might have tensile strengths in the 150 MPa range; the other materials are best considered as having near zero tensile strength, although compressive strength can be substantial.

Modulus: Varies from about 14 GPa for concrete to over 200 GPa for some ceramics.

Toughness: None of these materials could realistically be regarded as tough in the conventional sense, although some of the ceramic materials will be very scratch resistant. Castable ceramics and plaster are relatively easily scratched, as is graphite. None of the materials will be very resistant to impact, although if room temperature cure is used a concrete tool (with gel coat surface) can be robust due to a large thickness. What these materials will be resistant to is permanent deformation of the toolface. In general the toolface will either be as made (in dimensional terms) or broken in a very obvious way.

Tolerances: Ceramics and graphite can be machined to tight tolerances; the castables are limited by the accuracy of the master but these materials are rather less prone to dimensional changes on setting than other casting materials. The great advantage that this class of materials has is that of stability, dimensions are unlikely to vary.

Surface finish: The machined grades and some of the castables will take a very fine surface finish. Concrete will be very rough, but a thin skin of gel coat or GRP or a polymeric castable would normally be used as the actual toolface.

Longevity: As was noted above, the main danger is one of physical damage; in a production line environment it is likely that lifetime will be limited in this way rather than by 'wearing out'.

Tooling details: Integration of tooling details such as injector positions will generally be quite difficult; mechanical fasteners cannot easily be used and thermal mismatch stresses could be high; bonding with compliant adhesives is probably the best approach to attachment of mould furniture.

Coefficient of thermal expansion: Generally low, in the range of $7–12 \times 10^{-6}$/°C.

Thermal conductivity: In the range of 0.1% of aluminium for concrete to 10% of aluminium for some ceramics such as alumina.

Specific heat: Up to about 1000 J/kg/°C.

Maximum temperature: For most materials in excess of the cure temperatures used for RTM matrices; for concrete the thermal mass and difficulty of heating would generally limit uses to around room temperature, unless the mould were maintained at a constant temperature.

Gauge limitations: Fairly substantial thicknesses will be required to ensure adequate strength and handleability.

Acquisition time: The castable materials can be used to make tools very quickly; the machinable materials will be limited by machining time and any special machining requirements. In particular the number of vendors prepared to machine graphite is very low, both for reasons of cleaning up the cutting equipment and the requirement to isolate NC controllers etc. from conductive graphite dust.

Repair: Minor scratches etc. can be repaired in the usual ways; major breaks may be repairable as no plastic distortion of the broken faces will have occurred. This would be very unlikely to be acceptable in the long term but might allow a 'limp home' capability while new toolfaces are being made.

Modification: Machined systems should be able to be modified; cast systems are probably better remade.

Chemical compatibility: Problems would not be expected with chemical compatibility, all these materials will require surface sealants; the suppliers of tooling graphite also supply sealants and adhesives.

Size limitations: Machinable ceramics will not be available in large blocks; tooling graphite is available in large blocks and can be bonded up into larger pieces. The castable materials could be made in large pieces but the weight and handling difficulties may become excessive.

Limits to complexity: The limit is essentially the lack of strength and toughness in thin sections of any of these materials; they could be used in multipart tooling.

Graphite tooling might find applications where its very high levels of stability lead to the most reliable and reproducible tolerance levels; the same is true of machinable ceramics. Castable ceramics have some interesting properties in terms of high geometrical stability at low cost, but the lack of toughness is a real problem. Concrete's sole major advantage compared to the other materials is cost, as a mass cast material behind a thin gel coat or GRP skin its cost is substantially lower than any other material. For prototype activities, where the scale of the tool would lead to very high costs for alternative mass cast materials, the use of cast concrete can be an acceptable alternative.

One of the first RTM tools I ever made used concrete in this way to form the external collet of a complex seven-part tool; in this case the concrete was contained inside a steel tube to take all the tensile loads. The entire tool was built up from cast elements and completed in a few days at very low cost. Although it would have been woefully inadequate as a production tool it enabled prototypes to be made for 'touch and feel' evaluation without the expenditure of much time or money.

4.2.7 Aluminium

Aluminium tools can be produced by machining from billets or by casting to near net shape and final machining; the material properties and other factors may vary slightly between the two cases and both will be considered here.

Specific gravity: 2.7

Strength: 50 MPa (pure aluminium) to 500 MPa (strong alloy); for tooling work a medium range alloy would usually be used.

Modulus: 71 GPa

Toughness: Aluminium is quite a soft metal; it is easily scratched, dented and otherwise marked. Machined aluminium is all but impossible to destroy in normal use, but tool surfaces can easily be damaged beyond the point of economic repair by operator abuse or poor design that results in excessive force having to be used. Ejector design is therefore important as is the provision of suitable, non-scratching, scrapers and other clean-up tools.

Tolerances: Tolerances can be very good on aluminium tools. If castings are used they should be stress-relieved prior to machining; it is also a good practice to stress relieve before final cuts on toolfaces that are made from billet stock if a lot of material has been removed. Well stress-relieved materials should not suffer from permanent changes in dimensions thereafter as a result of thermal cycling, although problems can sometimes occur if both steel and aluminium elements are used within one tool. No problems will be experienced in achieving 'O' ring sealing tolerances on machined toolfaces.

Surface finish: Aluminium can be polished to an excellent surface finish. Maintaining that surface finish without any scratches or marks on the tool requires a great deal of care and attention to details of demoulding and cleaning practices. Cast aluminium may have some porosity that reduces surface quality somewhat. This porosity needs to be filled with a good sealant prior to release agent application.

Longevity: An aluminium tool is unlikely to wear out as a result of use, but the surface may be damaged and any repolishing to remove scratches etc. will slowly degrade the dimensions from nominal. An aluminium tool is more likely to be scrapped by a single act of carelessness (e.g. closing the mould with a demoulding tool in the cavity) than by a general reduction in quality.

Tooling details: Integration of other tooling details is generally very simple with metal tools. Any bolt holes in the aluminium should be lined with helicoil inserts and bolts should not 'bottom out' in the threads as differential thermal effects between steel bolts and aluminium can lead to marking of the mould surface. Large washers should be used beneath bolt heads, and for preference the use of bolted connections into the toolfaces should be avoided entirely.

Coefficient of thermal expansion: $23 \times 10^{-6}/°C$. This is high compared to GRP and especially CFRP. Problems can occur when aluminium tools cool down around CFRP parts leading to high loads being applied to the CFRP. The high CTE can be used to good effect when aluminium is used as an inner plug, for example in constant cross-section tubulars, when its high CTE leads to easy demould. Having said this, I have used aluminium tools to mould BMI resins via RTM with cure temperatures up to 200°C. As much depends on details of the mould geometry and operating practices (e.g. hot demoulding or mould opening) as on the material properties themselves.

Thermal conductivity: 200 $Wm^{-1}K^{-1}$. This very high value of conductivity makes it easy to ensure uniform temperature across the tool. Aluminium tools represent the standard by which others are measured in terms of ease of heating by a wide variety of means.

Specific heat: 913 J/kg/°C.

Maximum temperature: Above the cure temperature of RTM matrices.

Gauge limitations: Small aluminium tools that have no additional framing will commonly be around a minimum of 10 mm thick to resist the loads applied, cast moulds may have to be thicker for minimum casting gauge reasons. Thicker toolfaces might also be used to ensure the minimum point-to-point temperature variation.

Acquisition time: If CAD data are available in the right format for the generation of NC cutter paths the machining of aluminium toolfaces can be very rapid as the material machines very freely. Cast toolfaces will need some normalization or stress relief prior to machining. This and the requirement for patternmaking will lead to longer times for tool acquisition by this route. Several weeks may be required to acquire cast aluminium toolfaces and the necessary support structures.

Repair: Minor scratches can easily be polished out, but continual re-polishing of the toolfaces will eventually degrade the tool cavity dimensions. Major repairs can be effected by machining away the damaged section, inserting a new block of material and remachining the toolface. It is prudent to ensure that tooling data in three axes are available on the mould so that the tool can be set up easily on a mill for any repair work; this is true for all machined toolfaces.

Modification: Tools can be extensively modified in the same way as they can be repaired; major repairs or modifications will tend to reduce the tool's useful life somewhat.

Chemical compatibility: No problems would be expected with solvents or release agent compatibility, cast tools will need to have any surface porosity sealed or filled before use. The acid catalysts used with cold cure phenolics would be expected to cause corrosion. Damp storage conditions can also cause corrosion of toolfaces.

Size limitations: Limits are set largely by the size of the machine tools available to manufacture the toolfaces. There is also a limit to the size of solid billet of aluminium available. If a great deal of material removal is required it may be better to follow the near net cast route to minimize machining time. In principle tools could be sectionalized to overcome size limitations, but this would be expected to cause tolerancing and sealing problems.

Limits to complexity: Very complex tools can be made, sharp or thin edges can be easily damaged but, with care, complex moulds can be kept in good condition. Tools can be more complex than would realistically be achievable with GRP or CFRP.

Aluminium moulds have many attractive features for both prototyping and production work and have widely been used for both applications. Very good and reliable results can be obtained from such moulds but care has to be taken to avoid scratches to the toolface and to avoid the problems that can arise when steel and aluminium are used together. It is best to restrict the use of metal scrapers etc. in the cleaning up of aluminium moulds and provide plastic scrapers for shop-floor use.

4.2.8 Steel

Steel has some excellent properties as a tooling material and is normally used in the thermoplastic injection moulding industry and for SMC production. These are both high volume processes where the moulds are handled automatically; it is much less common to use steel tools for RTM.

Specific gravity: 7.9
Strength: minimum 300 MPa
Modulus: 210 GPa
Toughness: Steel tools, even if made from mild steel rather than tool steel, are much more robust in use than most other materials, being difficult to scratch, dent or break. Tool steel moulds can give extraordinary performance, but at a very high price.
Tolerances: Steel tools can be machined to very fine tolerances, and if properly made and used will not deviate from these tolerances with use.
Surface finish: A high polish can be achieved and retained on steel tools. Corrosion from damp atmospheres or fingerprints is the most likely reason for a fall-off in surface finish.
Longevity: Apart from major accidents the lifetime of a steel tool in RTM should be essentially infinite.
Tooling details: The steel is itself hard enough that inserts can be avoided, and bolts etc. do not have to have the special treatment given to aluminium toolfaces.
Coefficient of thermal expansion: 15×10^{-6}/°C.
Thermal conductivity: 30% of that for aluminium.
Maximum use temperature: Above RTM resin cure temperatures.
Specific heat: 420 J/kg/°C.
Gauge limitations: Steel toolfaces may be a little thinner than aluminium tools for strength and stiffness reasons, but the lower thermal conductivity may require higher thicknesses to ensure uniform mould temperatures.
Acquisition time: Steel tools will take longer to manufacture than aluminium tools because steel is not so easily machined.
Repair: Repair of a steel toolface is not likely to be required very often; minor scratches etc. can be polished out, more major damages can be treated as for aluminium. The very high forces required to damage steel tools may lead to a gross distortion of the toolface and the overall geometry should be checked before attempting any local repair.
Modification: Can be extensively modified if required.
Chemical compatibility: No problems expected with solvents. Acid catalysts would be a problem, giving rise to corrosion. The single most dangerous enemy of a mild steel tool is probably rust formation from storage in damp conditions.
Size limitations: Similar to those for aluminium; the increased tooling weight for a steel tool would make mechanical handling a necessity at a much smaller tool size than for most other materials.

Limits to complexity: Very thin or sharp edges on mould blocks can be used with more confidence than with softer materials; very finely detailed tools have been made from steel. In some cases the main danger has been the thin edges cutting operators rather than being damaged themselves.

Steel tools have many excellent properties, but the weight and hence slower thermal response are negative factors as is the high costs of manufacture. Steel moulds would only be likely to be considered for small and very complex tools, where tolerances are absolutely critical, or if production volumes are very high.

4.2.9 Cast iron

Cast iron is not an obvious tooling material in many ways, the general impression of such materials in many people's minds is of a coarse and brittle material rather than a good engineering material. Cast iron does, however have some very good properties which ensure it a place for some tools. Another material that has been developed as a cast metal tooling material is cast bronze (copper/aluminium/iron) this is much stronger than cast iron but otherwise has similar properties.[5] Recently, invar (an iron/nickel alloy) has been added to the list of metals from which cast tools have been made. Invar's prime claim to prominence as a tooling material is its extraordinarily low thermal expansion coefficient of 0.9×10^{-6}/°C, making for great stability in use. If anything invar has too low a thermal expansion coefficient even when compared to that of carbon fibre composites. The comments made below about cast-iron tooling apply in the most part to cast-bronze and cast-invar tooling as well.

Specific gravity: 7.2
Strength: 100–200 MPa
Modulus: About 150 GPa, dependent on grade
Toughness: Cast iron gives very hard and scratch resistant toolfaces; in the gauges likely to be used in practice, cast iron has adequate strength to resist all but the roughest handling without damage.
Tolerances: Machined surfaces can be very closely toleranced; properly normalized cast iron is a very stable material and dimensions would not be expected to change with time. Manufacturers of toolmaking accessories have taken advantage of this fact and offer standard size blocks of well-normalized cast iron as off-the-shelf items. Blocks such as these are excellent starting points for small-cavity machined tools as all faces are accurately machined flat and perpendicular to each other, making setting up very easy. These blocks will also machine more freely than equivalent steel blocks.

Surface finish: While not quite as good as machined aluminium or steel, excellent surface finish can be achieved.

Longevity: In service experience with cast-iron moulds has been very favourable and excellent longevity can be achieved.

Tooling details: Integration of tooling details is generally very good; the material is hard enough that most features will not require any inserts. Minor thread loads can be taken directly; heavy bolting loads might be better avoided.

Coefficient of thermal expansion: $11 \times 10^{-6}/^{\circ}C$, this is less than that for steel and half that for aluminium, about equal to that of GRP and around twice that of quasi-isotropic CFRP in plane.

Thermal conductivity: 35% of that for aluminium. To ensure good temperature uniformity a network of heater ducts or channels can be cast into the back face of the tool so that heating fluid can be circulated, giving excellent uniformity of temperature.

Maximum use temperature: Above the cure temperature of RTM resins.

Specific heat: 500 J/kg/°C.

Gauge limitations: Wall thickness is likely to depend on the skill of the foundry, and is unlikely to be much below about 15 mm, stiffening webs can be cast onto the mould face during the casting operation eliminating the need for additional external stiffening of the toolface.

Acquisition time: Small tools machined from stock plate can be acquired quickly; tools that must be cast will take perhaps 10 weeks for the pattern manufacture, casting, normalization and machining. The most critical factor is probably the choice of foundry as a high-quality, pore-free casting is absolutely critical to the manufacture of a high-quality tool.

Repair: Minor scratches etc. are easily dealt with, more extensive damage can be handled in the same way as for steel or aluminium tools.

Modification: As for other metallic tools.

Chemical compatibility: As machined the surface may be slightly porous, pore filling and careful 'running in' of cast-iron tools is required. As for steel, mould surfaces should be protected from damp environments, although some cast-iron grades rust less than steel, acid catalysts would also be expected to be a problem.

Size limitations: This is largely dependent on the initial stage of casting, although in principle sectionalized tools could be made. Cast-iron tools will be very heavy compared to some other tooling materials, but if use is made of external cast-in webs the total weight can be minimized.

Limits to complexity: Very thin edges or sharp corners will be more prone to damage in cast iron than in steel; small inserts can be made in steel to remove this difficulty. Cast-iron tools have been used to make some of the most complex RTM components currently in aircraft.

Overall, cast iron has a suite of properties that make it a very useful material for RTM moulds if the weight can be tolerated. The ability to avoid a lot of machining and finishing operations by near net casting and casting in back-face stiffening webs can make cast-iron moulds less costly than wholly machined metal tools, and for more complex shapes they can compete on costs with composite tools. The number of vendors that can supply the highest quality of castings is likely to be low for cast iron and even lower for other materials such as cast bronze or invar. This supply situation may act as a limitation on the specification or use of such materials. Tool designers wishing to take advantage of the benefits of these materials would be well advised to check the supply position at an early stage, especially if the tool procurement timescales are tight.

4.3 REQUIREMENTS FOR THE DESIGN OF RTM TOOLS

4.3.1 Introduction

In a study of the design of RTM tools the best place to start is with a consideration of what functions are required. As before, the basic assumption made will be that the tooling is rigid, rather than having some flexible elements.

- The tool has to form a cavity of accurate and reliable dimensions.
- It must be possible to introduce resin into this cavity in such a way that the reinforcement filling the cavity can be uniformly wetted out with resin. Excess resin must be able to exit the cavity and the tool cavity may have to be evacuated.
- The flow of resin from the tool must be controlled to keep the workplace clean and some element of mould sealing is usually required.
- Fibre packing and injection pressures must be reacted without excessive deflections or any damage to the mould structure.
- The heat required to cure the moulding must be provided and for very rapid curing systems it may be necessary to provide cooling to limit peak exotherm temperatures.
- For tools that are to be thermally cycled both heating and cooling may be required to gain the necessary control over the cycle time.
- The lay-up of reinforcement or preforms into the tool must be facilitated and the ejection of cured parts must be possible without damage to the tool or component.
- All of the above must be achieved reliably, with an adequate lifetime and be capable of good control and quality.

The tools are only part of the total production engineering setup and must be integrated into it, which may influence aspects of the tool design.

For example, in aerospace it is usual to separate clean and dirty processes. This means that the tool should be loaded in a clean area, while injection and clean-up/preparation take place elsewhere. In this case the tools must be moved from one place to another and this handling and transport requirement can affect all elements of the tool design.

The design of the RTM tool should be seen as one element of a total system design that encompasses component design, process design, production engineering and production line design, quality considerations, ease of use and ergonomics and operator training. For the best solutions these factors need to be considered in parallel so as to avoid any single set of requirements dominating the process and leading to non-optimum system solutions.

As a simple example, assuming a baseline case of a relatively high Vf% of around 50%, increasing target Vf% will dramatically increase mould closure loads; it is likely to make mould loading more difficult; it will make the mould closure stage more sensitive to defects or inconsistencies in preforms and make achieving a perfect mould seal more difficult; it will make the resin injection process more protracted and more sensitive to easy flow paths. The net result of this would be to increase labour costs, extend the mould cycle and associated costs and perhaps to increase the likelihood of defective parts. If the increase in Vf% occurs early in the product development cycle it is possible that the system could be modified to meet the changed requirements without too many problems. The worst case would be a desire to pack in more material as a result of components failing in validation testing, without changing the tool cavity dimensions, which is often impossible. Minor, or even major, changes to the design of an autoclave moulded component can often be made at any time without any great effect on the tooling or production route. This is not usually the case for RTM and it is essential that the correct design methodology is adopted for both components and tools etc. The critical factor is the involvement, in the product design process, of all those having an input into the manufacturing of the components.

4.3.2 Loads on the toolfaces

Loads arise from the pressure required to compact the reinforcement to the required Vf% and from the pressure of the resin injected into the tool, there may also be additional loads generated by changes in temperature leading to expansion of the resin. These loads have to be reacted in some way, either directly by the toolfaces or by some support structures or a press. The loads have to be reacted within deflection limits that are set by the tolerances on the part and by considerations of sealing tolerances to prevent leakage of resin or air. The reinforcement compaction pressure is unlikely to exceed 2–3 bar and is usually much lower. The resin injection pressure depends on the type of machinery used, essentially on the maximum pressure available when the mould becomes filled with resin and the bulk flow stalls. This can be from 1 bar or less up to tens of atmospheres in some pumped systems. It is becoming increasingly common for pumped systems to incorporate pressure relief systems to keep maximum pressures down to a few atmospheres. In some cases a slight flexibility of toolfaces means that the correct quantity of resin can be injected before any resin shows at the vents. In this case resin flow can be stopped before the tool becomes hydraulically tight avoiding the overpressure pulse associated with stalling the pumps. As the tool relaxes the excess resin will be driven to the vent positions.

When the resin is heated to cure (or exothermic heating raises the temperature) the resin will expand. If moulds are not sealed during cure the injection pressure will decay to zero and the resin expansion may drive resin from the tool; this can lead to some problems and can generate voidage as the resin shrinks. On the other hand only the reinforcement compaction pressure will be relevant to questions of tolerances, although damage due to gross overpressure from pump stalling or loss of sealing tolerances would still be a possibility. If moulds are sealed during cure some very slight flexibility in the toolfaces can be an advantage. Firstly, it permits a very small reservoir of resin to take up any resin cure shrinkage. If tools are essentially totally inflexible the decay in resin pressure as the resin cures and shrinks will be much more rapid, which can lead to shrinkage-induced voidage or poor surface quality (see Figure 4.1).

Secondly, the resin expansion on heating will generate an additional pressure, the magnitude of this will be related to the stiffness of the tool, being lower in a flexible tool. One has to be careful with this as flexing of the toolface across any stiffening ribs could lead to localized overloading which could give rise to fatigue problems. In conclusion, in some cases a controlled element of mould flexibility may give better results than total inflexibility, so long as the maximum possible pressure on the tool is known and controlled.

The worst case is when the tool has a large flat area such that all the loads are reacted by bending. For a square plate with fixed edges the centre point deflection is given by:[6]

$$y = \mathrm{A}.P.B^4/E.t^3 \tag{4.1}$$

Where A is a constant = 0.0138 for this case
P is the total applied pressure after injection is complete
B is the panel width
E is the toolface modulus
t is the toolface thickness

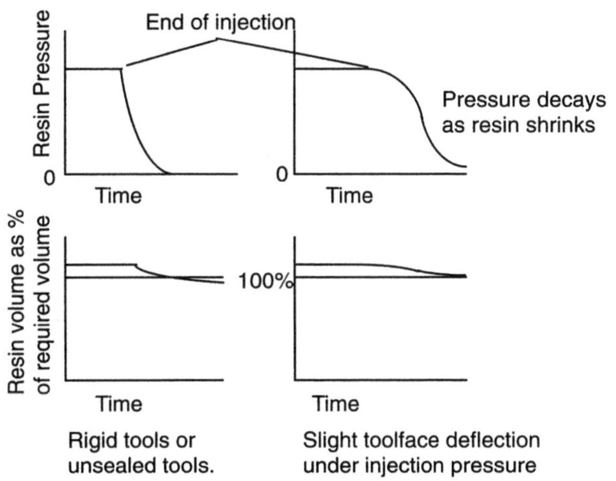

Figure 4.1 Resin shrinkage influences the pressure within the cavity during cure. For rigid or unsealed tools pressure decays rapidly after injection ceases, due to loss of resin or shrinkage. Additional shrinkage can lead to voidage. Where toolface deflection creates a reservoir of excess resin at the end of injection this voidage may be avoided.

Then for a given deflection,

$$t = (A.P.B^4/E.y)^{1/3} \tag{4.2}$$

The panel width has the greatest single influence on the required thickness. If B is set at 0.5 m, P at 2 bar (0.2 MPa), E at 70 GPa (i.e. an aluminium tool) and y at 0.05 mm (half the tolerance band for 2 mm ±0.1 mm), then t = 36.7 mm. In this case the total load to be reacted would be about 5000 kg. A steel toolface would be 25 mm thick, a GRP toolface around 60 mm thick and a CFRP toolface around 40 mm thick. If the actual tool were 60 cm square to allow for seals one toolface would weigh 36 kg in aluminium, 72 kg in steel, around 40 kg in GRP and around 22 kg in CFRP. Half the thickness tolerance band is assumed above, this would be taken out on both faces of the tool so that the edge dimension could be set at 1.95 mm and then the panel's maximum and minimum thicknesses would be 1.95 mm and 2.05 mm. A slightly lower edge thickness might be taken to allow for any slight variations in clamping effectiveness. Similar calculations to those shown above can be used to set the minimum spacing between clamp points to ensure the correct sealing tolerances.

If it is desired to use a thinner toolface for reasons of cost or ease of procurement then some bracing is required. If two steel hollow sections were used, each 7 cm wide, the effective panel size becomes 12 cm by

50 cm. In this case the total deflection is built up of the panel deflection plus the deflection of the bracing. Assuming that this is shared equally (i.e. 0.025 mm for each) other equations in reference 6 (Roark and Young, 1976) can be used to calculate the required thicknesses. This comes to 18.8 mm in aluminium, and to 14.8 mm if bracing is applied across as well as along the tool. A rough calculation indicates that for 7 cm hollow section steelwork a wall thickness in the region of 3.5 mm is needed which would weigh in at about 15 kg. The use of the steel supported aluminium toolfaces would only save about 6 kg, but would permit the use of thinner and more easily acquired aluminium sheet or GRP. Total tool weight would be in the region of 60 kg, compared to about 10 kg for an autoclave tool for the same duty.

As the tool area increases the unbraced thickness of the tool goes up rapidly (assuming loads and deflections as above), e.g. at 1 m by 1 m the thickness of aluminium required is 92.4 mm and at 2 m by 2 m the thickness is 232.8 mm. The tool weight in the latter case would be in the region of 3000 kg.

These examples are rather artificial as flat, large area, components are not the most obvious candidates for RTM unless the production requirement is high; in which case the tool is likely to be press-mounted with the press stiffness dominating the total deflection. Having said this, I have seen examples of flat tooling in which deflections are so high that they are unusable without substantial additional stiffening, which can be hard to arrange once the tool has been completed. A particular problem can be seen with flat and parallel-sided box structures such as can be seen in Figure 4.2; in this case tool deflection can produce an effective undercut making demoulding next to impossible from simple tools.

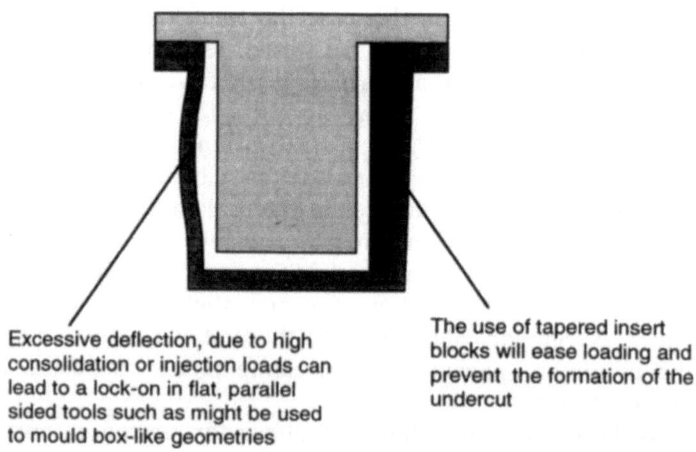

Figure 4.2 Tool deflections can lead to demoulding problems in some cases.

Some tooling materials, such as CFRP or nickel electroforms, are only going to be available as thin shells; the design of suitable stiffening arrangements for these is a major part of the tool design task. In general, toolfaces must be stiff and packing and injection pressure must be no higher than absolutely necessary, which indicates that these factors must be known to the tool designer at the earliest possible stage and not increased thereafter. Stiffening toolfaces with cheap steel sections is often more economic than increasing the average toolface thickness. The steelwork can also be thermally isolated from the toolface material to limit its influence on thermal cycling. In theory the tool geometry could be adjusted such that the tool has the correct geometry only when loaded with reinforcement, this ought to be the minimum weight solution but would have its own difficulties.

For more complex geometries the problems of deflection control are generally less severe, but the question of deflections and tolerances under load should never be ignored.

4.3.3 Thermal considerations

There are two major considerations, firstly the heating and cooling cycle that must be imposed on the tool. Secondly ensuring that the tool produces components of the correct geometry, taking into account differential expansions. The second point is considered in the next section.

For tools that are used at room temperature or some constant temperature, the heating strategy and the tool's thermal mass are not too critical. High thermal mass or low thermal conductivity tools may have advantages for constant temperature tools in that they lose little heat when opened for cleaning and loading. For tools that must be thermally cycled the thermal response is very important and must be considered carefully.

For the example considered above weights were estimated for toolfaces of various tooling materials with and without steel bracing. These were 36 kg and 14.5 kg for aluminium, 70 kg and 30 kg for steel, 40 kg and 17 kg for GRP and 26 kg and 10 kg for CFRP. Assuming a measure of thermal isolation of the support frames the thermal mass of the braced tools would be less than half that of the monolithic tools. As the toolfaces may have very different thermal expansion from that of steel the thermal insulation also serves the purpose of allowing movement between the tool elements without attracting loads into them.

If the tool cycle is 100°C (e.g. from 20°C to 120°C or 50°C to 150°C) and the thinnest toolfaces are used with the steelwork isolated such that its temperature is unchanged, then the energy requirement to raise the temperature of a single toolface is given by:

$$\text{energy} = \text{specific heat} \times \text{weight} \times \text{temperature differential} \qquad (4.3)$$

for aluminium	energy = 1323 kJ
for steel	energy = 1260 kJ
for GRP	energy = 1700 kJ
for CFRP	energy = 1000 kJ

All the above figures assume zero heat loss from the tools.

To define heating power requirements the target cycle time must also be known. Assuming that a 20-minute heating time is required the power requirements become:

for aluminium	power = 1.1 kW
for steel	power = 1.05 kW
for GRP	power = 1.42 kW
for CFRP	power = 0.83 kW

Again this assumes no heat loss and averages the power requirement over the full heating period; in reality more powerful heaters would be specified to deal with losses.

The question that now arises is: how will the materials react to having this power applied to them? The metals have relatively high thermal conductivities such that the temperature gradient across the wall of the tool will be fairly low; this is not the case for the composites. Calculations of transient heat transfer is a specialized field and will not be considered here; some simple calculations can however be made. In the very best case the whole surface of the tool can be considered as being uniformly heated, so that the power/square metre is four times that shown above. In steady-state conditions it is easy to calculate the temperature difference across the toolface corresponding to the power requirements noted above.

for aluminium	temperature differential = 0.33 °C
for steel	temperature differential = 0.77 °C
for GRP	temperature differential = 136 °C
for CFRP	temperature differential = 61 °C

In the instantaneous heating case, rather than under steady-state conditions, inputting the levels of power quoted above would lead to much higher temperatures around the heaters and very rapid degradation of the composite tooling. It is clear that low thermal conductivity systems, such as composites, mass cast systems or ceramics, are very limited in respect of thermal cycling rates and that the low-temperature variants are very susceptible to damage caused by power controlled heaters such as electric bar or plate heaters. If temperature cycling were a requirement the use of fluid circulation heating would be preferred to electric heating as the maximum temperature could be controlled. If temperature control at a fixed temperature is required then electric heating is a possibility, using low power heaters to limit overheating effects. Laminated in heater blankets have been reported and used.[7] One variant uses a thin skin of reinforcement to form the toolface, with the heaters above that and a backing of insulating material and further reinforcement to complete the toolface. This approach greatly reduces thermal mass and would make the most thermally responsive composite toolface, at least as far as the heat-up cycle is concerned; no provision is made for mould cooling. Using a similar technique with a mesh of fine pipes might be a possible solution to the heating and cooling cycle, but would be likely to present more practical difficulties. The recently developed space fabrics can also be used to generate a cavity through which fluids can be circulated to heat and cool the tool. Whatever approach is taken, the heating and cooling response, the control of heating and the thermal uniformity are likely to be much worse for low conductivity tooling materials than for metals.

In conclusion, when tools are dimensioned by deflections, the use of the necessary large thicknesses of non-metallic materials makes the design of tools that must be rapidly thermally cycled very difficult. If slow heating rates such as might be achieved with oven heating are acceptable, or essentially constant temperature is required or the toolfaces are very thin such as might be required for vacuum only RTM then non-metallic materials become more acceptable. Outside aerospace, temperature cycling of tools is rather uncommon and GRP and mass-cast tools are commonly used. If temperature cycling is a requirement the use of metallic tools can reduce the total number of tools required to achieve a given output and can thus reduce the tooling bill, even if each tool is more costly.

4.3.4 Thermally induced dimensional changes

When a transient heat flux is applied to one side of a plate, that side expands, causing the plate as a whole to bow. While this effect might occur during the heating of a tool the following notes will assume that the temperature of the tool is uniform throughout.

For the simplest case of a flat laminate, the example will be taken of a CFRP plate required to be 5 mm thick at 20°C, which is cured at 170°C. The thermal expansion coefficient of the CFRP in the thickness direction is assumed to be 30.10^{-6} °C. On cooling to room temperature the thickness will reduce by 0.0225 mm, thus the mould cavity dimension should ideally be 5.0225 mm at 170°C. If the tool were of steel its thickness at room temperature would have to be 5.0112 mm. In reality it is most unlikely that accuracies of this level would ever be required; the quoted thickness change is only about 20% of the minimum likely tolerance on a 5 mm thickness, and mould deflections under load would be a more serious problem than allowing for the CFRP's thickness expansion.

For the same laminate, assuming that it is required to be 1 m square and quasi-isotropic, at 170°C it will be about 0.6 mm longer than at room temperature. If the tool were made of aluminium it would have expanded by nearly 3.5 mm at the same temperature. If the tool were of the net shape design and cooled to room temperature prior to being opened the laminate would be exposed to considerable compressive strain from the tooling. If the resin were kept under pressure up to and during cure there would be a resin-filled gap between the tool and the layup (as the layup which fitted the tool at RT would be undersize at the cure temperature) leading to a trimming requirement even though the tool was designed as net shape. In principle the relative expansions could be accommodated by the use of a strip of very high-expansion coefficient material around the periphery of the tool. A quick calculation suggests that a strip of rubber about 16 mm wide would serve the purpose, but it is not known whether this approach to accommodating differences in thermal expansions has ever been tried in practice. If there were compelling reasons for the use of aluminium as a tooling material the best approach would probably be to use a shallow taper at the tool edge (beyond the end of the reinforcement pack). The mould would be opened slightly before cooling so that the laminate would be lifted out on the resin-filled taper as the tool cooled; the laminate would then require trimming. If the component were not flat but contained some moulding features at the periphery of the tool, it would almost certainly be necessary to allow for the differences in expansion coefficient between tool and moulding if aluminium were to be used for the tool. In this case the use of aluminium would clearly not be ideal and lower expansion coefficient material would be required. Depending on production rate and thermal cycling requirements, and the detailed shape, this might be a metal such as invar or cast iron or a composite material or graphite.

For a slightly more complex CFRP component shape such as two plates joined by a constant radius a change in shape of the component would be expected on cooling. This is because the in-plane and out-of-plane expansion coefficients are very different.[8] To a first approximation, the angular changes between the two plates can be estimated by $R/dR = A/dA$, where R is the initial thickness and dR is the through thickness thermal shrinkage, A is the angle and dA is the change in angle. This predicts an angular change of about 0.5° on a 90° bend, which is not a great deal but can sometimes be important enough to justify an allowance being made in the tooling dimensions.

As complexity increases these thermally induced geometry changes still occur but it becomes very difficult to account for them as the changes in geometry are constrained by the structure's shape and a combination of changes in shape and locked-in stresses is developed. In principle, 3D thermal FEM could be used to model these effects and thus produce tools that will generate the correct geometry at room temperature. In practice this seems to be very seldom attempted. The thermally induced stresses can be of more significance than any geometry changes, especially in the more complex components that are typical of advanced RTM. Components made from resins such as polyesters can show similar effects due to resin shrinkage rather than thermal expansion effects.

Having said all the above it is most common to manufacture tools to the nominal component tolerances and this is usually acceptable. Problems can be expected with large or complex components that are made in thermally cycled tooling with very different expansion coefficients. For smaller components it is often the case that the problems are not insurmountable and that the most expensive tooling options can be avoided. I have successfully used complex aluminium tools with CFRP and a cure temperature of 200 °C, which is hardly recommended on theoretical grounds.

4.3.5 Mould cavity geometry

The generation of the correct mould cavity geometry for ease of loading, avoidance of reinforcement trapping on tool closure, avoidance of high Vf% spots or easy paths requires more than the simple replication of the component drawing geometry. The process is made more difficult where high Vf% is required and/or components are of complex geometry and tooling is rigid. Therefore this section is primarily devoted to these components and simpler or low Vf% components require less attention to details. One reason for these potential problems is that rigid tool RTM moulding is much more sensitive to local fibre packing than other forms of moulding.

For example if the component is comprised of four plies and the plies have overlapped joints within them (e.g. because of ply size limitations) some areas of the layup will be of five or more ply thicknesses (see Figure 4.3).

If the tool were made to a constant mould gap the local Vf% could vary considerably. For example, if baseline Vf% is 50 for four plies it rises to 62.5% for five plies and 75% for six plies. This would lead to very much higher than expected mould closure forces and grossly distort the resin flow front shape. In open moulding techniques the ply count would simply show up as a thickness change in the moulding; this is not an option for rigid tool RTM unless the local changes in ply count are replicated on the tool's surface. The need to accommodate local ply count changes will decline greatly as the laminate thickness increases to the point where such changes cause only minor changes in Vf%. Even if the tool were modified to accommodate local ply count variations the layup accuracy would have to be higher than is commonly the case, to ensure

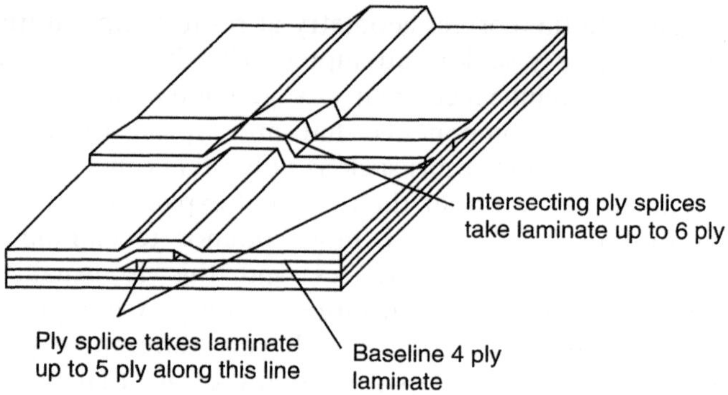

Figure 4.3 Overlapped ply splices in thin components should be avoided, or tooling modified to suit.

that the plies did not extend beyond the end of the tooling features that have been designed to take them. In general, if ply splices are required within individual plies it is preferable to utilize butt joints as this eliminates all the potential problems outlined above.

Even when butt joints are used, similar problems can arise with changes in section thickness, such as ply drops (again mainly for thin components). If the ply drop extends beyond its required position local Vf% and tool closure forces will, once more, be increased. As before, ply drop positions need to be replicated in the toolface. In addition, the layup tolerances must be all be taken on the negative side of the ply drop position to ensure that the ply cannot extend too far (see Figure 4.4).

Taking all the layup tolerance on the negative side can lead to the generation of resin-rich zones, unless layup tolerances are very good. When a series of ply drops is used to create a taper on the moulding surface it is common practice in prepreg moulding to intersperse the ply drops through the structure. There are two reasons why this practice might not be ideal in RTM. The first is that any tolerance problems that lead to high mould closure forces – and thus might indicate the likelihood of moulding problems – will be very difficult to correct during mould loading. If the taper is created entirely on the surface of the layup any errors can easily be detected and corrected. The second is that much of the cost savings associated with RTM arise from the use of preforming techniques; it is much easier to make two simple preforms, one for the main structure and another for the additional taper than to make a single preform incorporating all the elements.

In addition to the factors noted above the processes of preforming and mould loading can lead to changes in the thickness of the reinforcement pack. When reinforcements are shaped around a single curvature radius,

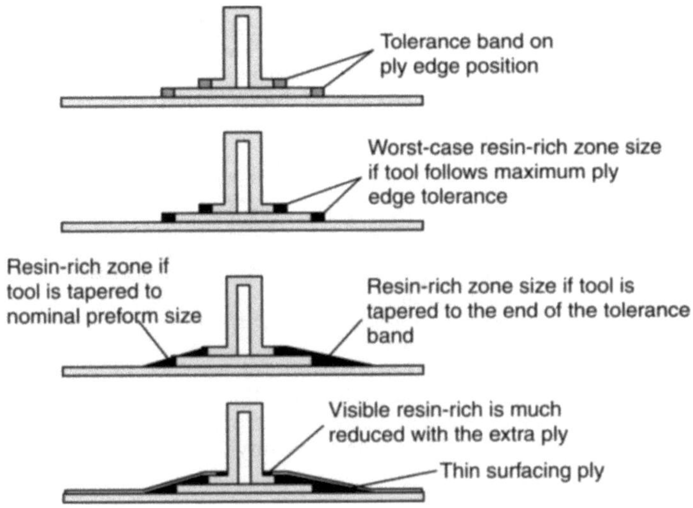

Figure 4.4 RTM with multi-ply preforms can lead to resin rich zones on the moulded surface; the size of these depends on preform tolerances, the fit to the tool cavity and whether surfacing tissue is used.

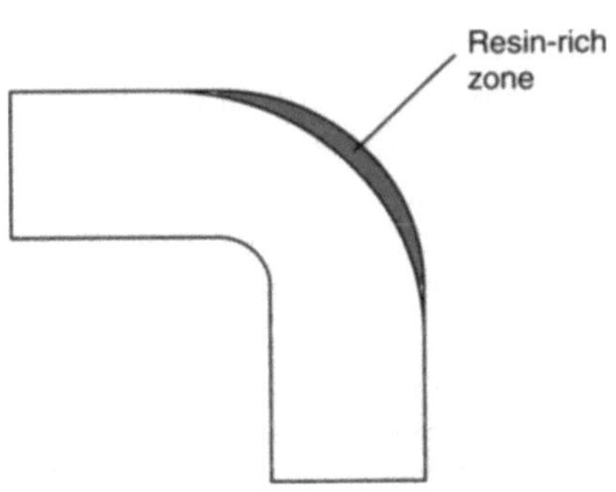

Figure 4.5 When cloth is formed around a tight radius it tends to become more compacted, leading to resin-rich zones and easy flow paths.

in a preform tool or by mould closure, there is a tendency for the reinforcement to 'cut the corner' as shown in Figure 4.5.

This arises because some of the forming loads may be carried in the plane of the reinforcement, acting as an additional compaction pressure around the radius and leading to a higher local Vf%. If the tool cavity is of constant thickness, resin-rich zones may be created which can be subject to thermal cracking and/or distort the shape of the resin flow

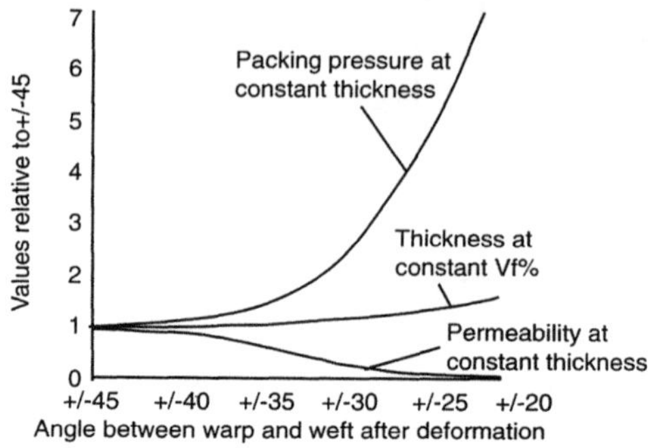

Figure 4.6 When woven coth is deformed the angle between warp and weft changes, leading to an increase in weight/unit area. If the thickness cannot increase packing pressure rises steeply and permeability falls steeply.

front. Either tool geometry modifications, changes to preforming or mould loading practices or layup modifications can be used to correct this problem.

When reinforcements are preformed into double curvatures there is generally a change in thickness, usually a reduction for mats and an increase for woven cloths.

Figure 4.6 shows the scale of the effects for woven cloth. The extent of the thickness changes obviously depends on the extent of the deformation. For example a woven cloth used to form a hemisphere is deformed close to its limiting deformation, so that the local thickness will rise by as much as 40% at four points around the periphery of the hemisphere. Predictive tools are available for woven cloths and could be used to determine local thicknesses and thus set tool geometry. For most reinforcement materials predictive tools are not available and if highly deformed preforms are used it would be prudent to measure the point-to-point thickness variations in the preforms. From this data a decision can be made as to whether the tool cavity or preform needs to be modified. Many components do not use highly deformed double curvature preforms or are of low Vf%, or the reinforcement thickness changes are accommodated by foam cores, for these components less detailed design of the tool cavity and preform is required.

In addition to the changes in thickness that are required to form the preforms, if control of the process is lost wrinkles and folds will be generated. These defects will increase the local thickness by a factor of 2 to 3, or even more if folds and wrinkles are superimposed. From all that has been said earlier it is obvious that such defects must be rigorously avoided by a careful inspection of preforms prior to mould loading. If such defects are regularly generated and cannot be eliminated by improvement to the control of the preforming process it is probably best if the preform is redesigned, probably by splitting it down into smaller sections.

It is clear that the determination of the tool cavity dimensions may require more effort than a simple counting up of nominal ply thicknesses, especially if a constant and high Vf% is required in a complex tool. The problems can be limited by reducing the target Vf% below that available in simple flat laminates. This approach is limited by the fact that if the Vf% is too low the reinforcement may move under the influence of the resin flow, leading to a lower Vf% limit around 40 to 45% for woven cloth. The use of higher bulk factor material such as felts or mats within a layup can also serve to limit some of these problems. As noted earlier, the involvement of tooling and production disciplines in the component design process can help to prevent the adoption of designs that are unsuitable from a manufacturing standpoint.

Having set the tool-cavity dimensions so that the tool and the preform are an acceptably close fit, the next step is to ensure that the tool can be closed without damage to the preform or tool. It is most important to avoid reinforcements being trapped between mould parts when the tool is closed as this will lead to oversize dimensions, and may cause poor sealing and poor quality. (This does not apply to tools designed with a pinch edge, and is most serious for net-shape tools.) For the example of a foam-cored vane such as shown in Figure 4.7, when the tool closes the preform will be consolidated to its final thickness.

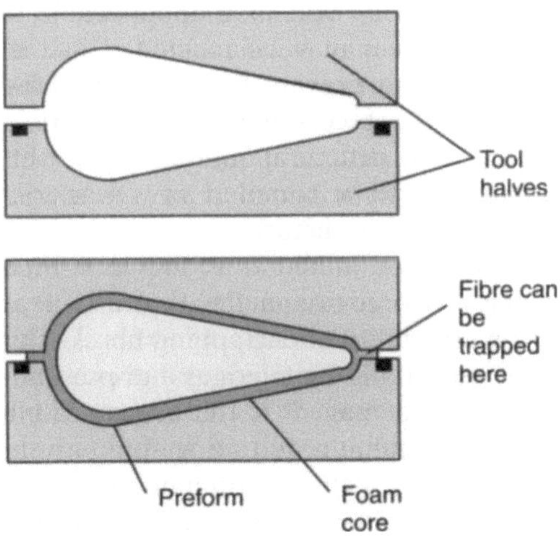

Figure 4.7 For some shapes, simple tooling may lead to problems with fibre trapping. If fibres are trapped the mould will not shut properly, dimensional control will be lost, trimming will be required and sealing may be impossible. If relief is provided to accommodate the trapped material sealing and dimensions will be right, but trimming is still needed and the moulding may be weakened.

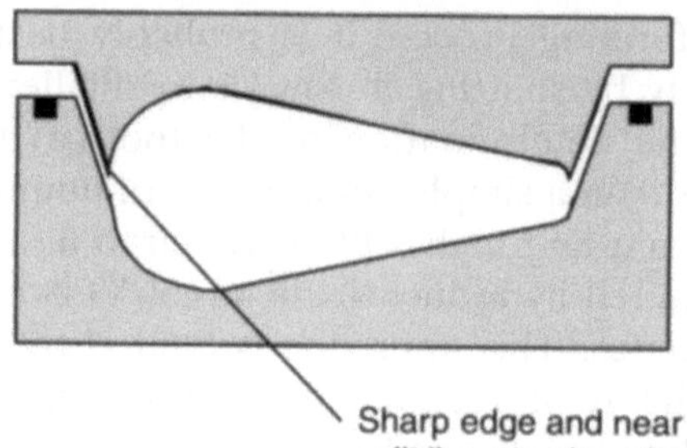

Figure 4.8 Problems with fibre pinching can be solved by more complex tool designs. Moulds as shown here are more costly than the simpler versions and the sharp edges are easily damaged.

In the simple tool shown, trapping of fibres between the tool halves is almost certain, leading to oversize parts and a trimming requirement. Assuming that this is unacceptable, several approaches can be defined to improve the situation.

Firstly, an expanding foam can be used. In this case the tool remains unchanged, but the preform is made slightly undersized. When the tool is heated the expansion of the foam generates the correct dimensions. This depends on the identification of a suitable expanding foam which may not be possible and is, of course, inapplicable to solid mouldings.

Secondly, a small recess may be machined in the tool such that the trapped material does not prevent the tool from closing fully. Trimming is still required and the recess may represent an easy flow path that distorts resin flow. For structural mouldings it would be necessary to ensure that the material to be trimmed away was 'sacrificial' and not part of the major load-bearing structure.

Thirdly, the tool can be modified as shown in Figure 4.8.

In this case the sharp edges on the tool and near vertical split lines greatly reduce the possibility of entrapping fibres. The edges of the upper tool part are now essentially knife edges and present an operator hazard, as well as being easily damaged. If the edges are blunted a witness will be generated on the moulding, requiring some finishing operations.

For some components, such as aeroengine vanes or other aerodynamic parts, the quality of the leading and trailing edges is paramount and it would be preferable to avoid any finishing operations in these areas. This can be achieved by the adoption of a tool design such as shown in Figure 4.9.

The removal of the split lines from the leading and trailing edges and the use of separate mould blocks ensures that mould loading is simple,

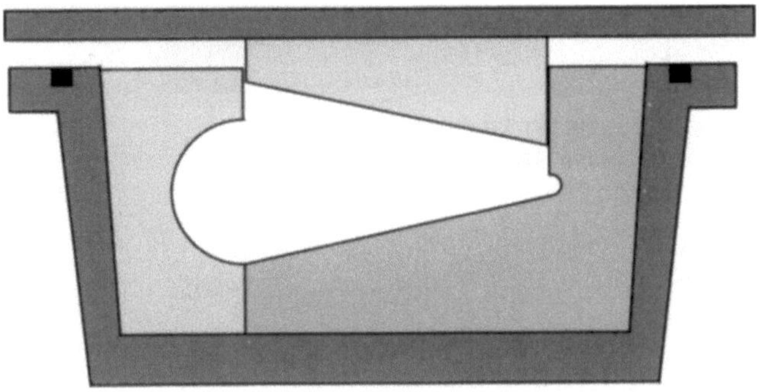

Figure 4.9 A mould such as that shown here can guarantee no reinforcement pinching and smooth surfaces without the problems of sharp edges; at an increased cost.

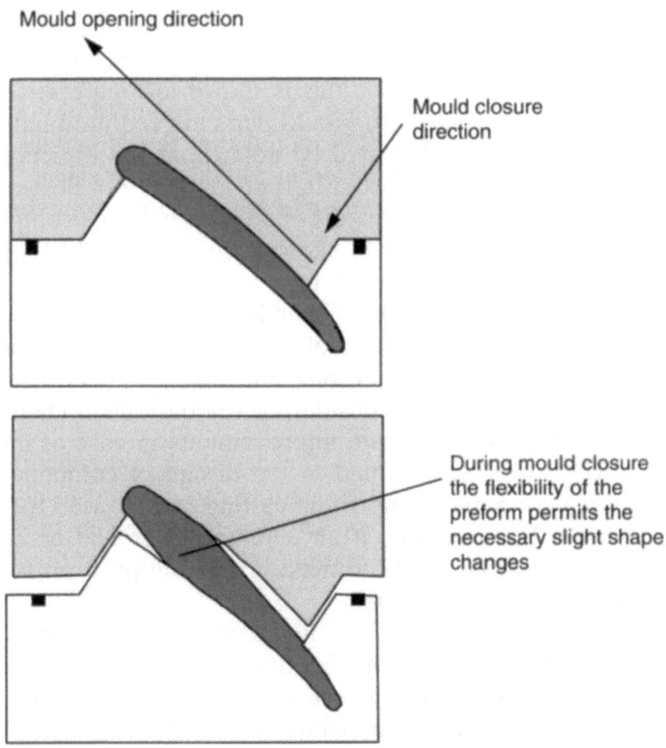

Figure 4.10 It is sometimes possible to design simple, two part, tools for shapes such as aerofoils that require the elimination of fibre trapping and no tooling marks at leading or trailing edges. Such tools would be more complex in operation.

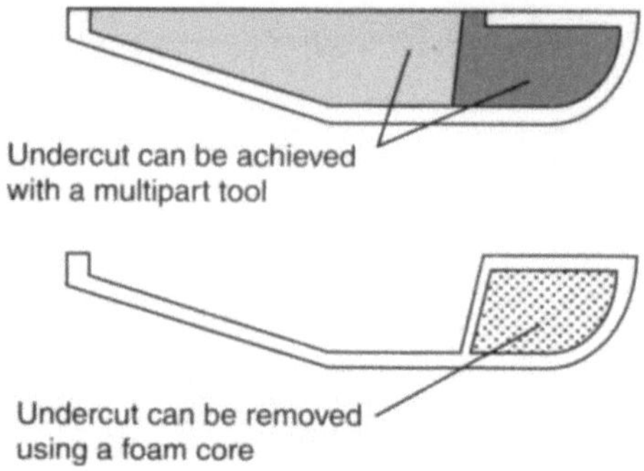

Figure 4.11 Undercuts that lead to complex tools can sometimes be removed by simple design changes.

fibre trapping is eliminated and operator hazards are removed. Technically this mould design may be ideal, but it has been achieved at the price of greater cost and complexity in the tooling. Another, simpler, design may sometimes be possible for such parts, depending on the detailed geometry requirements, this is shown in Figure 4.10.

In this case only two mould parts are required but the cavity geometry is still complex. It should be noted that the directions of mould closure and mould opening are different for this tool design.

Some geometrical features add greatly to the complexity of RTM tools; for example Figure 4.11 shows a complex RTM tool required to handle an undercut. If the undercut could be removed the tool design could be greatly simplified. For further information on how design modifications can lead to simplified tool design see the case study, Chapter 14. In general complex geometries can only be generated with complex tools and it is well worth considering whether small changes in geometry can lead to disproportionate improvements in ease of manufacture.

The experience gained in the design of components for manufacture by injection moulding or die casting can be used to illustrate such possibilities. For example, if an actuator lever had to be made it would be expected to be easier to make a reliable preform for a 'U'-shaped cross section than a solid one.

Figure 4.12 shows a box-like component in which a 'window' is formed through the component in what appears to be a direction perpendicular to the mould closure direction. This is achievable on two-part tooling by a small geometry modification from a flat-sided box. A moulded-in logo is also shown in Figure 4.12. Details such as this should be positioned parallel to the mould-closure direction; they should be shallow and the logo should be sunk into the tool. The position of any mould parting lines should be carefully considered and parting and flash lines should be kept

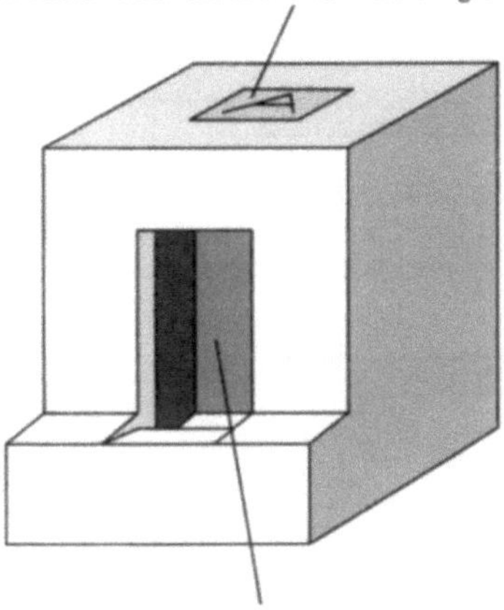

Figure 4.12 In this case a simple change to the part geometry allows a great simplification in the tooling.

away from flat areas as these are difficult to clean without marring the surface. If net-shape mouldings are being made they should connect to resin gates via thin flash gates to minimize clean-up costs. Many details and design possibilities have been explored in other tooling dominated manufacturing methods; the experience in these areas is worthy of some study.[9, 10]

Having discussed the generation of the correct cavity geometry it is necessary to move on to consider other elements of the overall tool design.

4.3.6 Mould seals

Seals are used to control the flow of resin out of tools and to permit evacuation of tools. When tools are not evacuated, such as is generally the case with polyester resins, the prime requirement is to control the resin flow and prevent it from contaminating the working area. This can be achieved by the use of a pinch zone or by the use of a mould seal. When tools are to be evacuated the tool must be sealed and the edge of the tool can either be a plain seal type or a net-shape type (see Figure 4.13).

The pinch zone works because the resistance to the flow of resin through reinforcement is a very strong function of the volume fraction

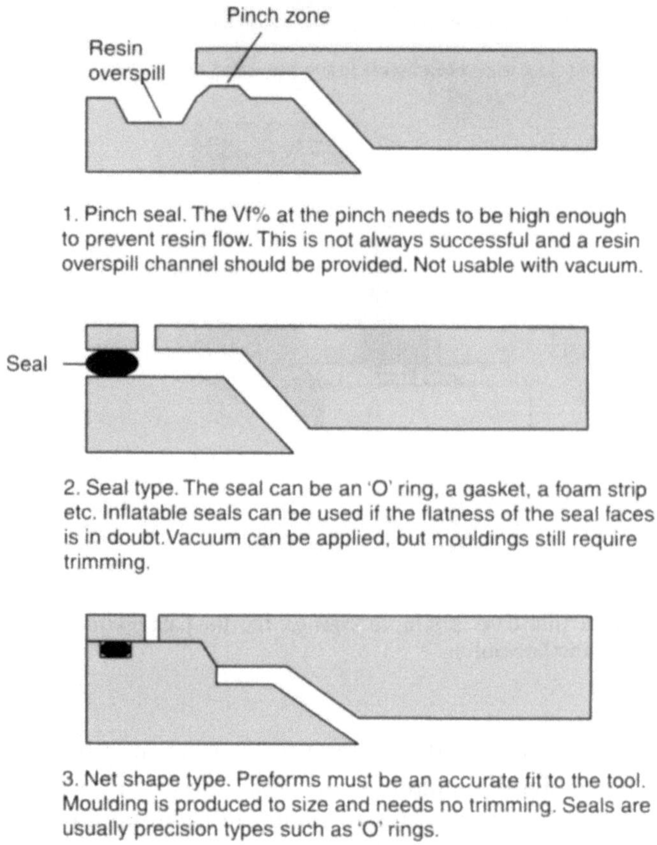

Figure 4.13 A variety of options are available for the edges of tools, three possibilities are shown here.

and thus of the compaction pressure. If the mould gap at the edge of the tool is reduced, a zone of high Vf% is created, which is the pinch zone. The intent is to create an area through which air can penetrate, but which resin cannot fully penetrate in the time available under injection pressure. The rise in back pressure as the resin reaches the pinch zone can in principle be used to signal the termination of the injection phase. Exact design of a pinch zone would require the knowledge of the reinforcement's permeability at high Vf% which is seldom available. Pinch zones tend to be designed by rule of thumb, say a 25% or 30% reduction in laminate thickness. In practice, pinch zones usually seem to fall short of the ideal and quite a lot of leakage of resin past the pinch zone can often be seen. This may be due to toolmaking tolerances, wear in this region of high clamping pressure, or inconsistencies in the reinforcement layup. When

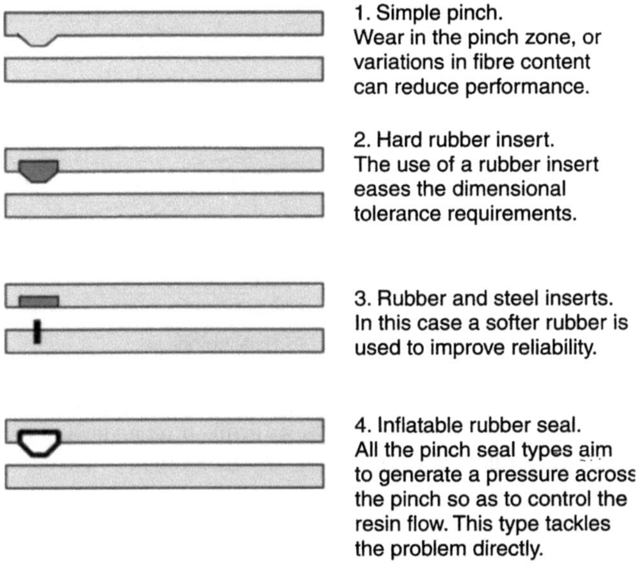

Figure 4.14 Even within the simple pinch edge tool there are a variety of design options. The costs will tend to increase from design 1 to design 4. It should be noted that designs 2, 3 and 4 could be modified after the tool is made to permit the use of peripheral injection or vacuum.

polyester resins are being used leakage past the pinch zone can lead to higher than ideal levels of atmospheric styrene. These problems can be minimized by incorporating some measure of flexibility into the pinch zone arrangements. This can be achieved in a variety of ways (see Figure 4.14).

It is axiomatic that mouldings made with pinch zones will require post-moulding trimming. Equally, since the toolfaces do not touch, other means must be found to control tool cavity dimensions. The use of cup and cone spacers permits the control of both mould gap and axial location. The use of pinch zone tools seems to be declining, presumably because of the problems noted above with styrene emissions and the maintenance of a clean working environment. I have seen many early RTM tools that leaked resin liberally onto the floor around them, as they lacked even a simple gallery to carry excess resin away from the leaking pinch zones. Such practices would no longer be acceptable.

Even for tools that are not evacuated it is now commonplace to use seals within the tool. Tolerancing requirements for seals that do no more than guide excess resin to a collection point are relatively low. Seal types include foam or solid rubber gaskets and various shapes of solid or hollow

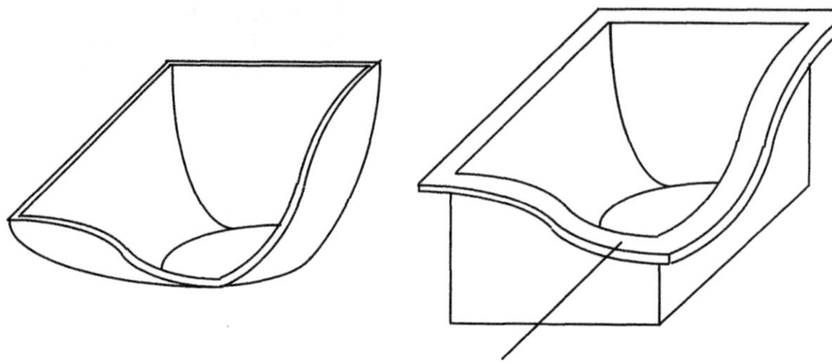

Figure 4.15 For components such as this, where the component edge does not fall on a flat plane, sealing the tool can be difficult. If high vacuum is not required the seal face can follow the contour of the component.

section rubber strips. These may be incorporated into seal grooves or used against flat faces. In the latter case, as the tool halves are not in contact, spacers are used to control tooling dimensions as for pinch zone tools. For this type of seal it is usual practice for the seal face to follow the contour of the edge of the moulding (see Figure 4.15).

The design of such seals is straightforward and presents no real difficulties, so long as the two halves of the tool are a reasonable match to each other.

Sealing tools that must be vacuum tight to a few mbar requires much closer seal tolerances and a better surface finish for the seals to seat against. Metallic tools seldom exhibit problems in this area. Problems may be seen in achieving the necessary flatness and surface finish in mass cast and composite tools. In this case softer and larger diameter seals tend to be used, rather than the small diameter 'O' rings that are often used with metal tools. There are some operational advantages in using small diameter 'O' rings in that if the tool seals against vacuum it is essentially guaranteed to be fully closed. It is much harder to achieve the required tolerances for vacuum seals when they follow the contour of the edge of the moulding than when they fall onto a flat plane. For this reason it may be convenient to sink the component parting line below the tool parting line (see Figure 4.16).

Tools like this may be made with either a trimming allowance on the edge of the moulding or can be made to give a net-shape part. Inflatable seals have also been used in RTM. They can be used where achieving sealing tolerances is difficult, and to replace the use of valves in fully closing off a cavity to allow curing under pressure. In this case the

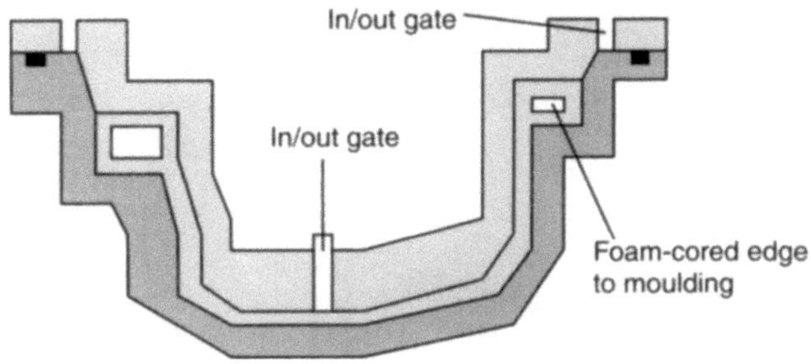

Figure 4.16 Net-shape moulds have many advantages when preform accuracy can be assured. They reduce trimming costs, permit the use of a flat seal line even if the component edge is not flat and are useful when the edge is foam-filled.

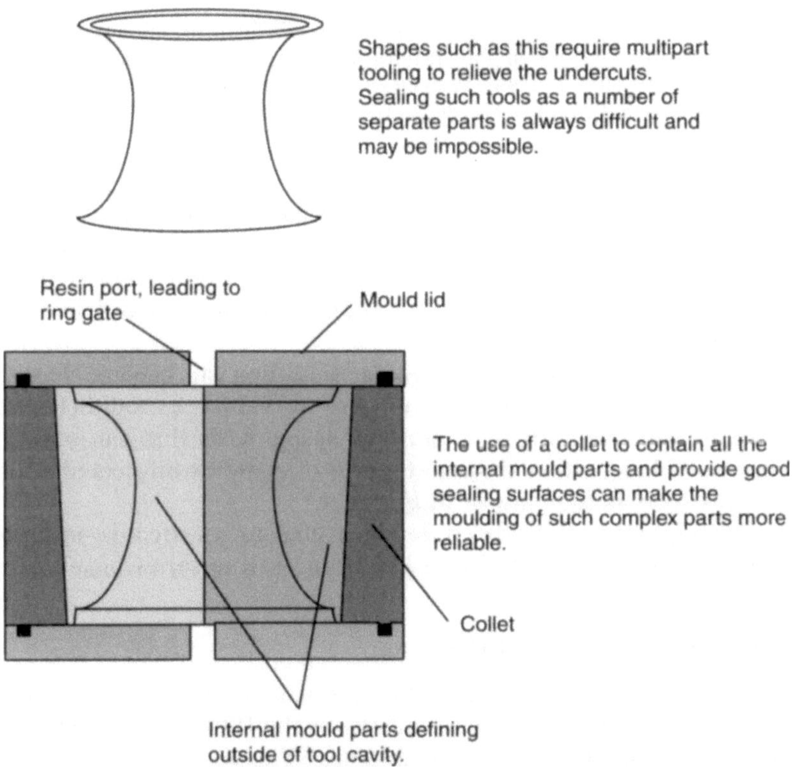

Figure 4.17 If the mould cavity is of complex shape and requires several sections to define the geometry, then sealing can be difficult. Collet moulds can make sealing much easier.

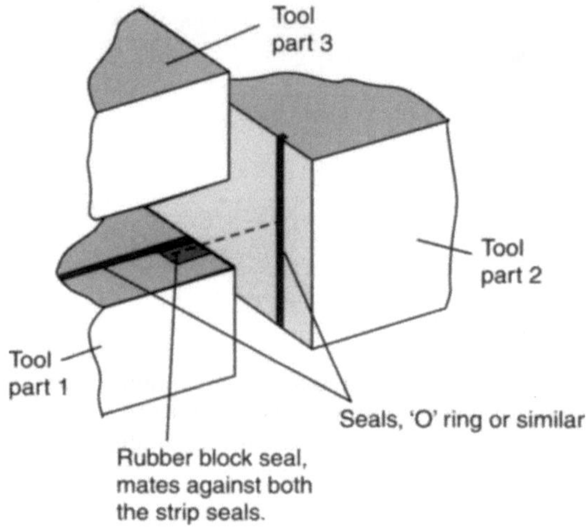

Figure 4.18 Complex, multipart tools are very difficult to seal against vacuum and resin pressure. The design shown here uses block and strip seals to effect a seal between three blocks. This has been demonstrated in development work, but in many cases collet moulds would be a simpler option.

inflatable seal is placed inboard of the resin out gate. They will also permit a simple, two-part tool to be injected under vacuum conditions while held slightly open. This reduces Vf% and thus injection time. At the end of injection the tool is fulled closed to eject excess resin and give very high quality mouldings.

If tools require more than two parts, sealing can become very difficult. The most common solution is to utilize a collet type mould (Figure 4.17).

As an alternative it is possible to design seals that can seal on more than one plane, either through the use of complex bifurcated seals or via designs such as that shown in Figure 4.18.

This type of seal has been demonstrated to be effective in prototypes, but it is not known whether it has been used in series production.

4.3.7 Mould closure and clamping

Consideration of these requirements starts to introduce questions of the production engineering requirements for the RTM production line as well as the design of the tools themselves. The requirements are that the tool should be open for cleaning and loading, the tool halves should then be brought together and closed against the resistance of the reinforcement and locked shut such that injection pressure does not cause excessive deformations or resin or vacuum leakage. As noted earlier the loads may be relatively high as a total thrust even if the pressure is kept to moderate levels. The total load can be estimated from a knowledge of the pressure required to consolidate the reinforcement and resist movement due to the flow of resin. In general the reinforcement clamping pressure needs to be of the same order as the resin pressure to prevent movement under flow conditions. The allowable deflections depend on considerations of sealing and component tolerances etc. Factors such as production rate and cycle or the need to separate clean and dirty areas can also influence the design of mould closure and clamping equipment.

Three stages have been identified above, large-scale mould movements that are outside the influence of the reinforcements, mould closure against the resistance of the reinforcement and locking the tool shut. Although all these stages can take place in one piece of equipment it is convenient to take them separately.

For light tools the first stage can be manual, simply lifting one tool half and placing it on the preform held in the other half: although this always involves a risk of dropping one toolface onto the other and is best avoided. As weight increases, manual handling becomes impossible and large-scale tool movements must be mechanized. At the simplest a chain hoist can be used to raise and lower the top tool half, pneumatic cylinders can also be used for this duty. The hoist mechanism should permit the rotation of the top tool half for tool cleaning and preparation. Some tools are difficult to open when filled with a cured moulding (because of their geometry). In this case it may be necessary to use higher loads for the initial opening, with the hoist mechanism just used to lift the tool clear. If high tool opening forces are experienced it is important that the tool cannot become misaligned as this may lead to jamming; to prevent this, several lifting points may be required. For some tools, such as those required to make parallel-sided tubes, special rigs may be required to separate the two tool halves. For the heaviest tools, or those that must be rapidly cycled, the tool opening can be accommodated within a press frame, preferably with a tilting upper platen for improved access to the top tool half. Such presses can also be used to close the tool and to resist injection pressure. If the mould cycle is short the tool normally remains permanently coupled to the press. If the cycle time is long there are advantages to using the press to close the tool and providing a separate tool locking mechanism, as this permits one press to serve several tools. This would also be a preferred option if clean and dirty activities must be rigorously separated.

Assuming that the tool lid has been lowered but not clamped shut and locked, what options are available for carrying out these actions? Two types of solution are possible. The first type combines the actions, the second merely provides the locking forces.

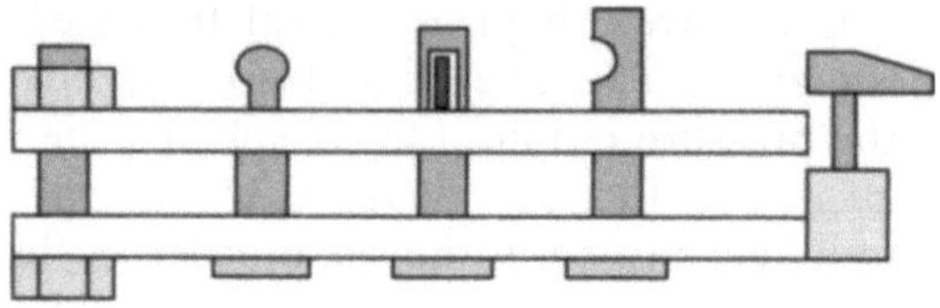

Figure 4.19 Various types of mechanisms can be used to clamp tools in the shut position.

A. Simple bolts can be an effective, low-cost solution at low production rates or long cycle times.
B. The ball end on this type would be picked up by a hydraulic actuator attached to the top tool face. A commercial version is available.
C. A simple locking bar holds this type closed. This would only really be suitable for use with compliant or inflatable seals.
D. A rotary can, attached to the top tool face would be used to hold this type in the shut position
E. Standard quarter turn tool clamps could also be used.

Types A, B and E could provide final mould closure forces as well as holding the tool shut. Types C and D would require separate facilities to close the tool.

Figure 4.19 shows various options. Simple bolts (A) can be used to clamp and lock the tool. For long-cycle-time tools, the time to fasten the bolts is not a major issue and the use of bolts will be cost-effective. For tools that must be rapidly clamped the use of hydraulic actuators can provide both closing and locking forces. Normal tool-holding clamps (E) can be used and other types are commercially available, specifically designed for RTM tools (B).[11] (In either case they should be thermally isolated from heated tools.) The question then arises as to how many such clamping and locking devices are required. The deflection of the tool edge must be controlled; this can be achieved by many clamps on a thin edge or by a single clamp on each edge operating through a stiffening structure. Stiffened tool edges make that element of the tool more costly, but may reduce total costs if hydraulic actuators are used. It generally seems to be the case that either a lot of bolts or a very few hydraulic actuators are used. One word of warning should be expressed here. If a stiff framework is used to collect the loads so that one or two clamp elements can be used, the loads can become very high and the load carrying capacity of the system elements must be carefully checked. I have seen clamping systems fail in service as heavy bolts sheared or weld lines tore apart. This is seldom dangerous, but can be distinctly embarrassing. If a press is used to close the tool it can either be used to resist injection pressure or separate locking devices can be used.

As before, bolts could be used, as could over-centre clamps. Figure 4.19 (C and D) shows two other possibilities that might find application in particular cases.

The details of the design in these areas depend as much on how the tool will be used and the production engineering environment as they do on the loads that will be experienced. It is therefore vital that these factors are known to the tool designer.

4.3.8 Ejection of mouldings

Having injected and cured the part in the tool and removed the top mould half, it would be ideal if one could reach in and simply lift out the component. In some cases, where the geometry is favourable, this is possible. Simple tools such as rubber suction cups that permit a flat or slightly curved surface to be gripped may also help in lifting out components that would otherwise be very hard to grip. In the general case, mouldings do not fall out of the tool and must be assisted by the adoption of some ejection strategy. In the absence of a designed-in ejection strategy the natural ingenuity of the shop-floor operators will be called into play and methods will be developed to extract the mouldings. These usually involve hammers, wooden wedges, screwdrivers, bottle jacks and so on, often used across the sealing surfaces of the tool or even in the tool cavity. It requires no great imagination to realize that such techniques are likely to degrade tools and may damage components. Equally, such techniques tend to provide an uneven lift to the component so that jamming in the tool is not unheard of. Lastly, such techniques can be more time-consuming than doing the job properly. It is therefore axiomatic that the design of a safe and easy to use ejection system should form part of the tool design process.

In addition to the shape of the component and tool, the production engineering route is important as is the tool-heating strategy. For example, access to the back of tools may be limited by the use of oil heating galleries or the use of bag presses to close the tool. In these cases the ejection must be accomplished from the front face of the tool. Whether ejection is carried out from the front or back face the basic requirements are the same. Highly localized forces are best avoided; ejector features that can jam or become inoperative due to resin leakage should be avoided; features that lead to extended cleanup time or require a lot of maintenance should be avoided. Extraction loads should be controlled such that components cannot twist or jam as a result of movement. If several ejectors are used it is best if their movements are coupled together to prevent jamming. In practice this means that displacement control is preferred to load control; if load control actuators such as simple hydraulics are to be used they can be linked together through a loading frame.

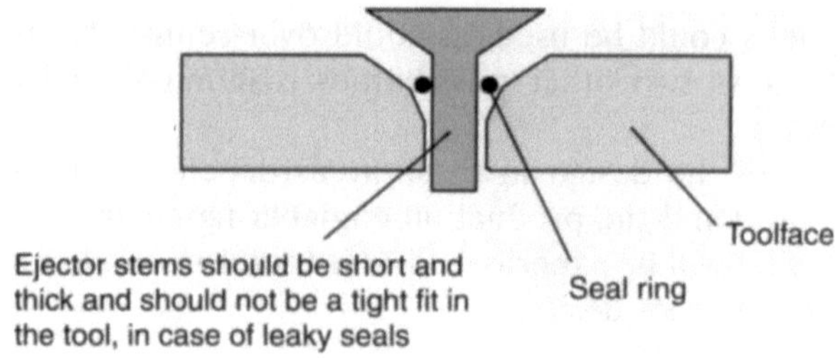

Figure 4.20 Careful design of ejection facilities is always required, the simplest form of ejector is shown here, this type is operated from the back of the tool. As an alternative the seals can be fitted to a groove in the ejector head, permitting a larger head to be used. This can be of particular advantage for thinner components.

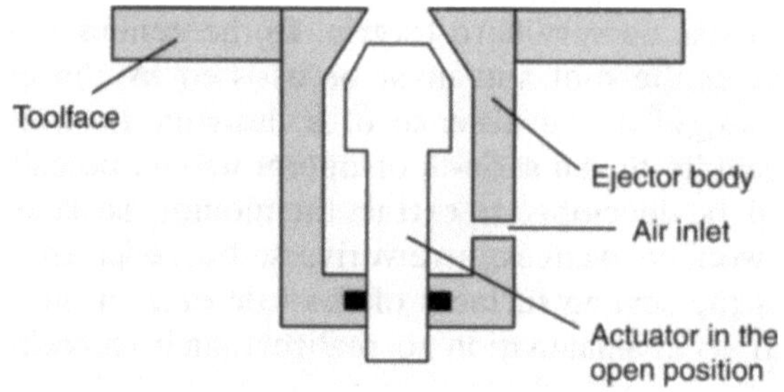

Figure 4.21 Air ejectors can be very useful to break the seal between tool and moulding, but do not provide full ejection.

Backface ejectors include conventional pin ejectors and 'big head' variants of these (Figure 4.20) as well as air ejectors (Figure 4.21). The latter type do not really provide full ejection, their function is to break the component away from the tool wall and other methods are needed to lift the part out. As noted earlier, problems can arise if some areas of the toolface cannot be used for ejectors because of the heating type, as this can lead to ejection from non-ideal positions. The ejector head diameter should be large enough that the localized ejection forces do not damage the part. For very thin mouldings the 'big head' type would be preferred. Pin ejectors normally show a witness mark on the finished component. This can sometimes be disguised by making the edge of the ejector coincide with a change in section in the tool.

Figure 4.22. shows a 'big head' type of ejector that might be used for a component of circular symmetry, in this case the ejector doubles as the resin in-gate and the edge of the ejector falls on a section change.

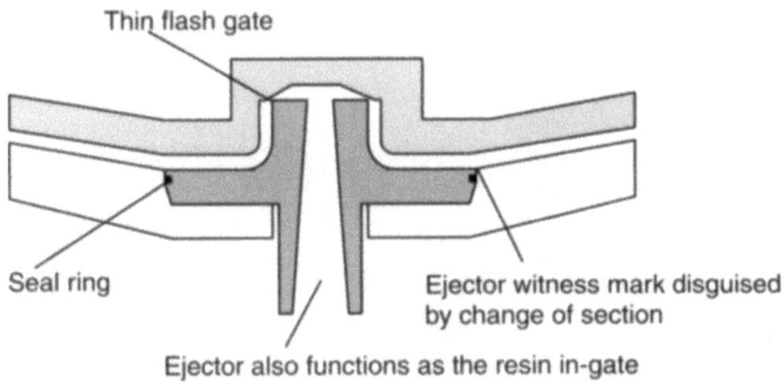

Figure 4.22 Ejectors can have other functions within the tool in addition to ejection.

Experience has shown that even when pin ejectors carry seals, some leakage of resin can sometimes still be seen. To guard against this causing problems the clearances between the ejector stem and the toolface should be large to prevent this area being held solid by cured resin. The length of the ejector stem should be held to a minimum to avoid buckling the stem. Lastly, for tools that must seal against a high vacuum, backface ejectors are less suitable as any vacuum leakage can only be rectified by removal of the preform.

Frontface ejectors require no access to the rear of the tool and can thus have advantages in many cases. The simplest version is the rubber

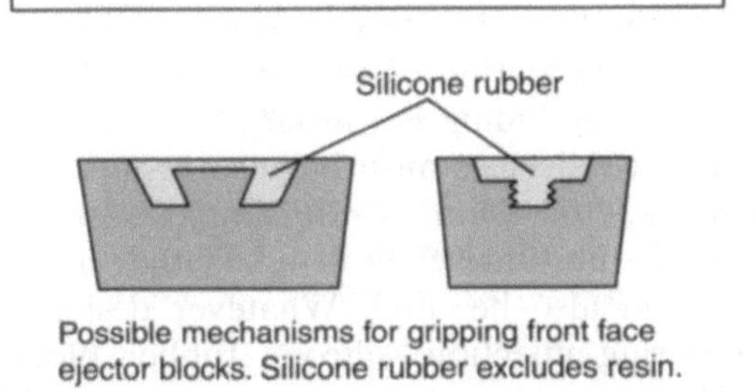

Figure 4.23 Ejectors can be operated from the front face of the tooling, this often has advantages.

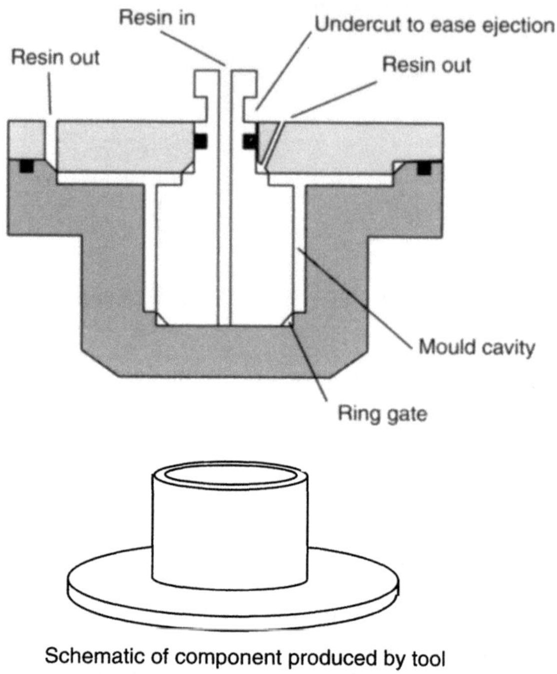

Figure 4.24 Loose blocks within the tool can be used to generate the required gating, provide ejection facilities and ease mould loading requirements.

suction cup noted earlier. These are primarily useful where the issue is one of gripping the moulding rather than high forces being required; several such suction cups can be used to ensure that extraction loads are uniform. When more load is required some form of loose block ejector in the mouldface is required. These can be sited on the periphery of the tool or within the area of the moulding if the geometry permits (Figure 4.23).

In some cases the loose blocks can form major elements of the tooling (Figure 4.24). A combination of suction cups and edge ejectors can also be used where the moulding is too large for edge ejection alone to be effective. Front-face ejectors must have some means available to impose the extraction loads on them. In the simplest case screw threads could be used, although these are slow in use. Other methods such as shown in Figure 4.23 can also be used. Whatever design is used, resin must be excluded during injection; silicone rubber blocks can be used to ensure this. The use of frontface ejectors via loose blocks increases the complexity and thus cost of the tooling, but has some compensating advantages such as mechanical simplicity and reduced possibilities for

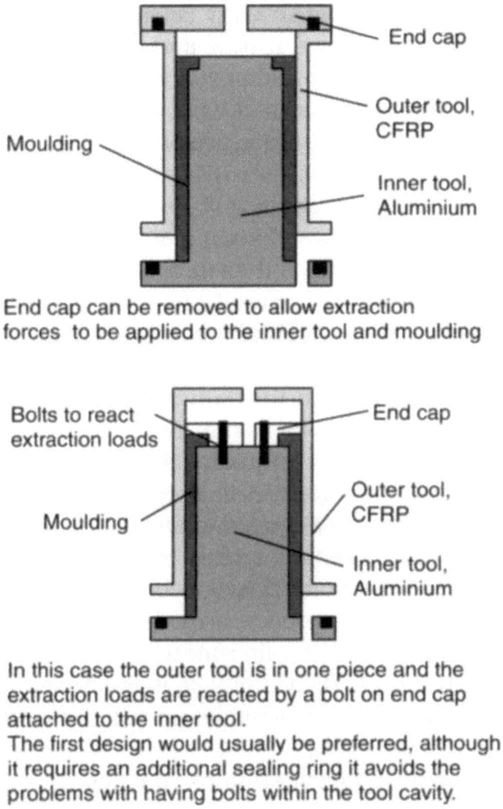

Figure 4.25 Absolutely straight-sided tubes need careful tool design. Two possible options are shown here.

leakage, failure or jamming, as well as avoiding the necessity for access to the back of the tool.

Some types of moulding require special treatment, these include moulds for parallel-sided tubes. Each case will be different, but the general principles of uniform loading and adequate extraction area will be the same in all cases. Figure 4.25 shows extraction options for a parallel-sided tube mould.

Paradoxically, some of the most complex components that must be made in collet moulds may require no extractors at all, as the mould parts fall away when the collet is opened. This not always the case and moulding extraction from collet moulds does need to be considered carefully. Melt-out or soluble tool elements can also be used to simplify moulding extraction. The tooling that makes these elements must be as carefully designed as any other mould tool.

Whatever type of ejection strategy is chosen it is commonly found that the first few mouldings from a new and complex tool are very difficult to extract easily. This is perhaps most common with tooling materials that have some porosity (e.g. cast metals) but any tool may have to be 'run in' carefully. Few things are more irksome than having a moulding stuck within a tool cavity, the only answer is patience and care in slowly extracting the moulding without any damage. In the last resort the tool is more valuable than the moulding and it is better to destroy the moulding during extraction and then reconsider the extraction strategy than to damage the tool. I know of one case where the first few mouldings to be extracted from a complex tool with many vertical rib elements took more than an hour to extract. By the tenth moulding the time was down to minutes and by the thirtieth the mouldings could be extracted more or less by hand. The tool had five ejector points, four peripheral and one in the centre of the moulding. The first few mouldings required the use of all ejectors and a great deal of patience. By the tenth only two of the peripheral ejectors were needed and by the thirtieth only the central ejector was required. The point here is that if observations of ejection requirements on production tools had been used to design the ejectors only one ejector would have been provided. In this case the mould could never have been run in as the first moulding would have remained stubbornly fixed in the tool. If in doubt it is much better to overdesign the ejection system to avoid problems in running in, or variations in ease of release during the tool's life.

Also of relevance to considerations of ejection is the choice of mould release agent. New tools need a careful application of mould sealant and release agent, but it should be noted that the tool in the example quoted above had been very carefully sealed and prepared as the geometry made problems likely. Several coats of sealer followed by the application of a film forming, bake on, release agent is generally the best practice. Dry types of release agent such as PTFE sprays may be rubbed off by preform movement during tool closure and are not ideal. The bake on release agents are generally supposed to be usable for multiple releases. Unless the moulding releases very easily it may be better not to rely on this and to use a light wipe-on coat of release agent between mouldings, especially in areas of complexity. Build-up of release agent should be avoided as this damages surface finish and may even degrade the release performance. The state of the release system should be monitored after each moulding and periodic stripping and re-application may be required.

To conclude, the extraction of the moulding from the tool is critically important and the design of extraction strategy is an important part of the tool design task. Extraction strategy influences the ease of use of the tool, the time required for moulding extraction and the time taken for mould cleaning and maintenance. Choice of the extraction strategy, therefore, has direct economic consequences on the production of components.

4.3.9 Integration of inserts into RTM mouldings

Inserts are often required in RTM mouldings and the integration of inserts into mouldings is often quoted as one of the advantages of RTM. Inserts can be metallic or non-metallic and can serve a variety of purposes. Typical inserts might be metal plates or threaded inserts, big-head type fasteners, tubes etc. Moulded in holes require similar approaches to inserts, and will also be considered here. The question of whether a moulded-in insert is preferable to a post-moulding operation cannot be addressed in a general way, as each possible application needs to be addressed on its own merits. In general, the advantages are that inserts can be precisely located relative to the rest of the moulding if moulded in and that post-moulding operations are avoided. The disadvantages are that the use of inserts will often make the tool more complex and can lead to mould loading or other difficulties. Lastly, if the component is cured at high temperatures or used over a wide temperature range thermal mismatch stresses can arise between the moulding and the inserts. Metallic inserts are probably more widely used in low-temperature cure GRP mouldings than in high-temperature cure CFRP as the thermal mismatch strains are much lower in this case. For example, only invar would be an adequate match to the CTE of unidirectional CFRP (and even then there would be a gross mismatch between the transverse CTE of the CFRP and the invar); titanium might be a reasonable choice for inserts in quasi-isotropic CFRP and steel and aluminium are increasingly poor choices for inserts in CFRP. For GRP steel would be an adequate match to the CTE, but aluminium would be unlikely to be used unless manufacture and use were to be at room temperature.

The simplest insert is a plate of metal enclosed within the layup. Such plates might be drilled and threaded post-moulding to carry loads into the structure. If several bolts are to be used in one area this is likely to be a better solution than the use of separate threaded inserts. Some sort of drill jig would be expected to be used to ensure the accuracy of the holes to be drilled. If the inserts are steel a magnet inserted into the toolface can be used to hold the insert in position during layup.

If the insert penetrates the surface of the laminate, or its position needs to be tightly controlled, a tooling feature must be provided to locate the insert. These usually consist of an insert carrier that is let into the surface of the tool and which is sealed against resin entry.

Figures 4.26 and 4.27 show some possible insert and carrier designs. If the insert is in the direction of mould opening it may not be necessary to penetrate the toolface completely in order to locate the insert carrier.

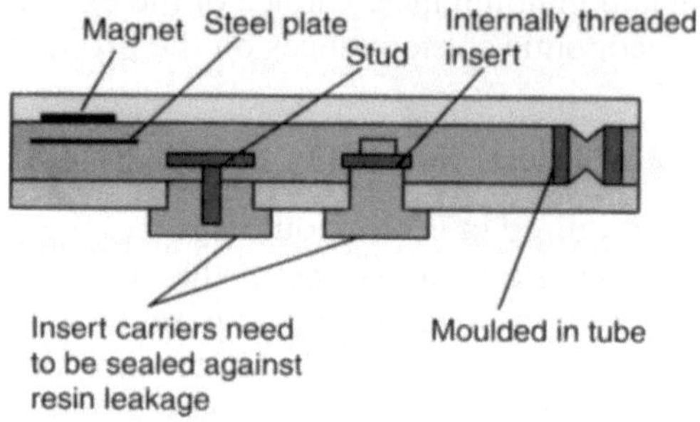

Figure 4.26 A wide variety of types of inserts can be utilized within RTM tools.

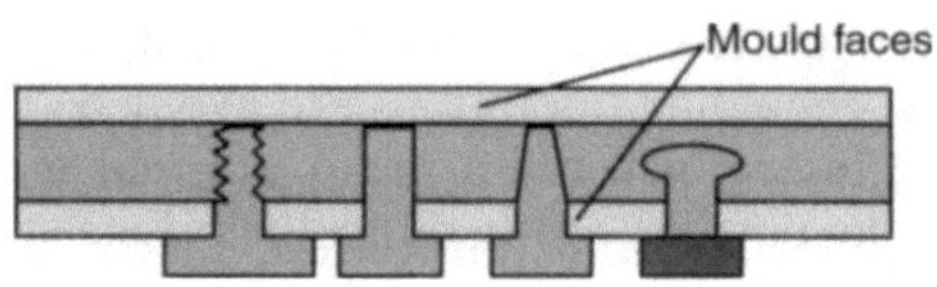

Figure 4.27 Holes can also be formed as 'inserts' within RTM mouldings. Tapered, plain, threaded and irregular holes can be made. Plain and thread forming inserts would be withdrawn before mould opening. Taper forming inserts need only be withdrawn if they are not parallel to the mould opening direction. Irregular holes would require a melt-out, or similar, former.

In many cases it is necessary to remove the insert carrier prior to tool opening in order to prevent damage to the tool or moulding. If inserts are used to give highly toleranced features in the moulding, it is axiomatic that layup accuracy will need to be just as good. Overall layup and tooling costs would be expected to be increased when inserts are used.

To decide whether inserts should be used, all the implications for costs and quality need to be carefully assessed. In addition to the obvious changes in labour and tooling costs (which need to include any additional costs if inserts are not used, e.g. drill jigs) a careful assessment needs to be made as to any changes in process risks and quality implications. For example, if a metallic attachment point were required in a CFRP moulding it could be moulded in as an insert, or a plain or stepped bore could be moulded in so as to minimize part-finishing operations (Figure 4.28).

Considerations of CTE mismatch might rule out the insert. For the moulded-in bore the positive features would be accuracy and reliability of position and the minimization of machining operations with their attendant costs and risks of damage to the moulding or machining inaccuracies.

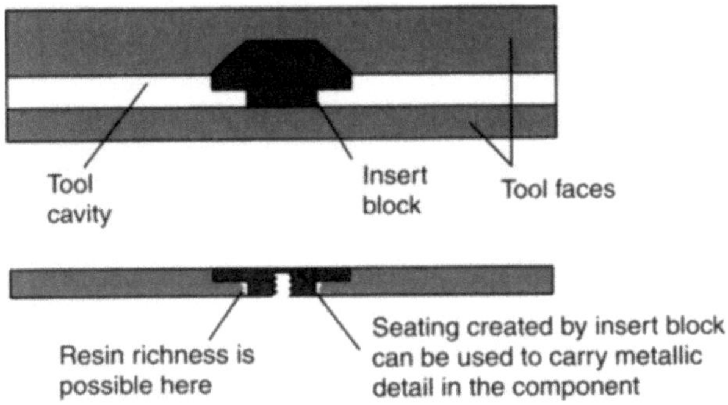

Figure 4.28 Inserts can be used to create a seating for post-moulding bonding operations.

The negative features would be higher tooling costs, an increase in the layup accuracy requirement, possibilities for the formation of resin rich zones around the bore and perhaps an increase in the possibilities for damage due to careless handling of the tool. All these factors would have to be weighed before a decision was made on which approach to take.

4.3.10 Resin distribution

This section covers the way that resin is distributed uniformly through the tool, from the resin in-gate position(s) to the resin out-gate position(s). Connecting these resin in and out gates to the resin supply equipment is dealt with in section 4.3.11.

The aim is to achieve a smooth and predictable mould filling through the reinforcement, without the flow front becoming distorted by variations in local Vf% or geometrical features and without air entrapment. At the same time it is desired that deflashing and cleanup operations are minimized and that the tooling be kept as simple as possible, and it may be necessary to carry the resin flow through a small patch of reinforcement to act as a QC sample. Lastly, it may be a requirement that the resin distribution system leads to a minimum injection time and or minimum resin wastage. The best solution to any single one of these requirements can easily lead away from the best solution to other of the requirements. All the requirements, therefore, need to be considered at the same time, taking into account geometry, fibre orientations, Vf% of the reinforcements and the tolerance on Vf%, and the possibilities for the generation of easy flow paths, which may be a function of the preform accuracy.

Some cases are very simple, for a circular part the resin could be introduced at the centre or the edge with radial resin flow. Injecting from the

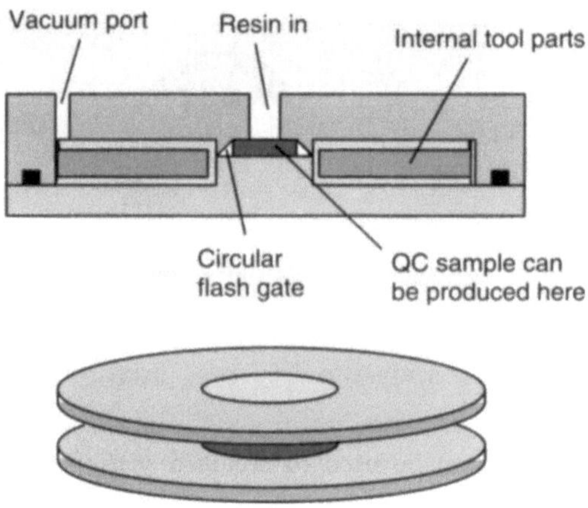

Figure 4.29 Thin flash gates can be an effective way of introducing resin to some components, as shown here. Components such as shown here are difficult to mould by autoclave methods. They have been successfully moulded by RTM with very small corner radii.

edge would be much more rapid (with constant pressure resin feed), but could not be used with a pinch-off tool, and problems could be experienced in an evacuated tool if the vacuum seals showed any leakage. If the part was a solid disc a surface port would be used at the centre point, if it was a ring a flash-gate could be used to minimize clean-up costs (see Figure 4.29).

Square or rectangular shapes would be almost as simple as circular shapes. Resin in-gates could be on the outside periphery with an-out gate at the geometric centre of the part, or the positions could be reversed. Alternatively, one edge of the component can be used as an in-gate and the opposite edge used as an out gate. This has the advantage of leaving no witness marks on the face of the component but there tends to be an easy flow path at the other edges of the tool that can lead to non-uniform flow and poor quality. For high aspect ratio rectangular parts that have a central in-gate the shape of the flow front will start out elliptical but when the flow front reaches the edges it can be deformed in the same way as for edge gated tools. If it was necessary that the resin be gated in from the centre of the tool it would be more usual to use a line gate along the long axis of the rectangle to ensure that the flow length was close to equal at all points. In this case a line gate on the surface might require substantial deflashing and the use of a loose block to generate a more easily cleaned thin flash gate might be considered (Figure 4.30) even though tooling costs are increased.

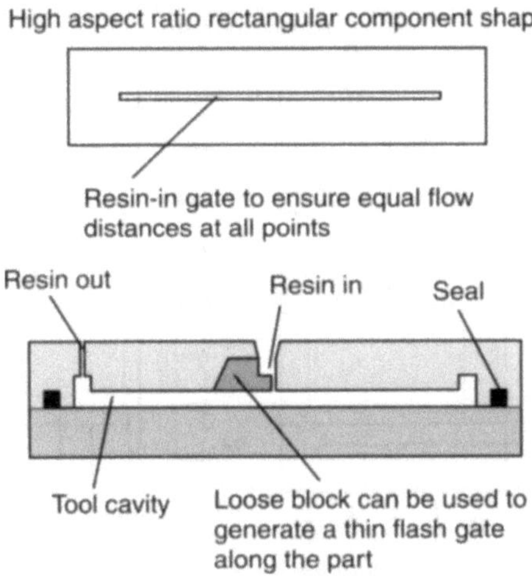

Figure 4.30 Loose blocks can be used to create a thin flash gate in a long thin part.

Even for such simple shapes as those considered above, the gating arrangements would require more care if the reinforcement permeability is grossly anisotropic, e.g. for UD reinforcement.

As the component geometry becomes more complex the correct positioning of the resin ports becomes increasingly difficult to predict, especially as complex geometries can also lead to easy resin paths. The problems are not so severe if tools can be reliably evacuated to very low absolute pressures prior to injection and the seals are guaranteed not to leak during injection. (Seal leakage is very important, the maintenance of a good vacuum by the use of a large vacuum pump that can overcome losses is not always an effective solution.) Even with evacuated tools in and out gates cannot be arbitrarily positioned, the vacuum acts to ease problems of uneven flow rather than eliminate them completely. It is necessary to use some predictive tools to make an estimate of how the resin will flow between the in and out gate positions. Various computer programs are available, some are developed specifically for RTM and others are general-purpose programs which were developed for thermoplastic injection moulding. Many of these programs deal poorly with easy flow paths, assuming a constant resistance to flow over all the component surface. Flow models have been developed that do permit the definition of easy flow paths but these are more complex to use than the simpler models.

In addition to computer-based methods, simple pen and paper methods

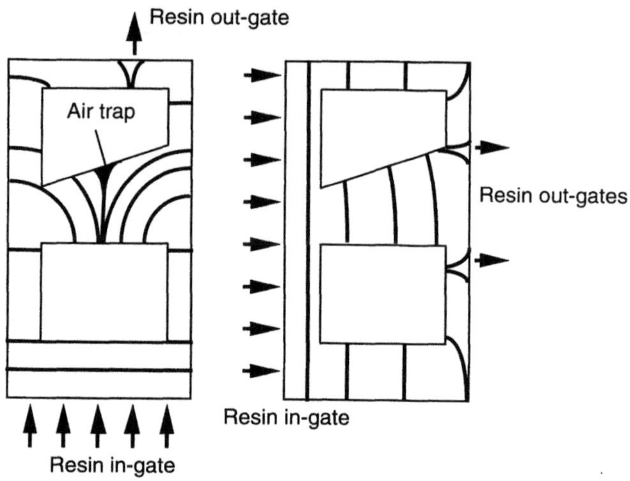

Figure 4.31 Simple pen and paper methods can be used to investigate gating options quickly.

Simple flow modelling 'rules'.
Equal flow in each time interval.
If flow front splits, proportion by relative cross-section at the point of splitting.
Expanding flows are of elliptical form.
Perturbed flows tend to revert to straight lines or simple curves.

can be used, based on a few simple rules such as assuming a constant flow rate with time and proportioning flows at splits in flow fronts.

Figure 4.31 shows the output of a simple pen and paper evaluation of a given shape. This shows that one potential injection direction would be expected to entrap air while another should give more uniform flow. The existence of easy flow paths can be taken into account in a crude way in the simple pen and paper models, for example by assuming that flow is twice as fast in these areas as in the bulk of the part.

Figure 4.32 shows the same part as previously, assuming easy paths all around the preform edges. With the easy paths, air entrapment is predicted with the gating arrangement that gave no air entrapment in the absence of easy paths. The pen and paper methods have many advantages. They take very little time to carry out such that many possibilities can be quickly assessed. They are conceptually simple and thus avoid the black-box nature of computer methods. They force tool designers to think about the best options rather than simply relying on computing power.

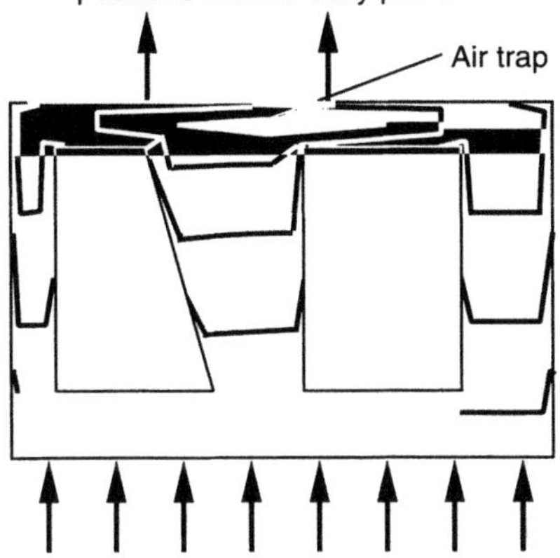

Figure 4.32 Simple pen and paper methods can also be used when the flow front is disturbed by easy flow paths. In the case shown here it is assumed that all preform edges act as easy flow paths and that resin flows down these at twice the rate through the reinforcement. When this assumption is made there is a strong probability of air entrapment in cases where uniform flow predicts even mould filling.

On the other hand, they will not be effective in predicting mould-filling time unless volume controlled injection machinery is used. Pen and paper methods are at the very least a useful method of scoping out the possibilities that need more intensive computer modelling.

If, as a result of the flow modelling exercise, no ideal (i.e. simply and easily achieved without adding to the tool's complexity or leading to high resin wastage) in and out gate positions can be identified there are various options for improving matters. It may be possible to modify the part geometry to smooth out the resin flow, e.g. where two split flow fronts rejoin. It may be possible to utilize a high level of in-mould vacuum, although this has the limitations noted earlier and the highest levels of vacuum cannot be used with polyesters. It may be possible to place the out-gate(s) in the positions of predicted air entrapment, although these positions may be sensitive to the assumptions made about easy flow paths and the part to part variations in preform accuracy. If some areas of the component are to be trimmed away prior to use it may be possible to ensure that any entrapped air is concentrated in these areas. Lastly, easy flow paths can be eliminated by very careful sizing and placement of the preforms. This can be achieved in a laboratory environment but is unlikely to be a stable solution in a production environment. In

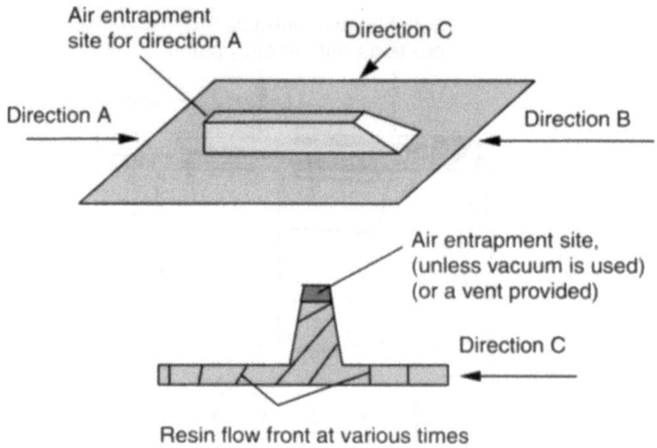

Figure 4.33 In ribbed mouldings air can become entrapped at the top of ribs. Injection directions A and C would be likely to entrap air. Injection direction B may avoid air entrapment if the taper angle is low. Air may still be trapped at the other end of the rib, especially for short ribs.

practice, a combination of the these approaches may be used, as may the use of gating options that add to the tooling's complexity. Such a gating option would be the use of secondary injection gates which are only operated as the resin front reaches some point in the tool to force a change in flow front shape.[12]

As an example of other options, Figure 4.33 shows a ribbed component with three potential injection directions. Directions A and C would be expected to lead to air entrapment. In direction B the part geometry has been modified in an attempt to provide a smooth resin flow into the rib and air entrapment might be avoided. In this case the best resin inlet position may be along the top of the rib; alternatively an air vent could be used if the position of air entrapment is reliable from moulding to moulding. Arranging the reinforcement such that the permeabilities in the rib were different to those in the rest of the moulding might also be useful in controlling the fill pattern.

For components to be made at a short cycle time the speed of injection is a critical factor and computer-based methods are usually needed to predict this. It is not necessarily the case that the best in-gate positions to minimize the injection time are the same as the best in-gate positions with respect to uniformity of flow and resistance to air entrapment. Equally, if speed of cycling is critical the use of features such as loose blocks to ensure that resin enters the preform at the correct place will not be ideal as this will increase mould loading and cleanup times.

In general it may be better in such cases to limit the complexity of the moulding to ensure that flow is predictable, even if this means not taking full advantage of the parts integration potential of RTM.

For quality critical components the fill quality is likely to be more critical than the fill time or overall cycle time and a wider range of options is available for ensuring that quality is adequate, e.g. by using loose blocks to carry resin into the required positions within the tool (Figures 4.24 and 4.30).

Lastly, the notes above refer to resin in and out gates. It may sometimes be possible to dispense with the latter and use unvented tools. In this case any air that is displaced from the preform by the resin flow is concentrated into a compression chamber within the tool. For this to be successful the tool would have to be well sealed and rigid and injected at a high pressure. In addition, either the in gate must be sealed during cure or gelation must follow rapidly at the end of injection to prevent the resin being forced back out of the tool. The advantages of such a design are that resin overspill and wastage are eliminated, cleanup time associated with the outgates and seal contamination is eliminated reducing total cycle time, and the pressurized compression chamber ensures that the resin is cured under pressure which generally improves surface finish and quality. There would be substantial costs associated with manufacture of adequate tools for this approach, making it unlikely to be followed except for relatively small items required in very large volumes.

4.3.11 Injection machinery and its interfacing with the tool

The previous section raised the question of the choice of correct position(s) for the introduction of resin into the tool. This section is concerned with the interface between the injection machinery to be used and the tool into which it is to be injected. It is necessary to start this with a brief description of the available types of injection machinery.

There are two basic types of injector, the first meters and mixes two or more components and provides this under pressure to the injection point.[13] The second uses single part or pre-mixed resin and simply provides the injection forces. In the simplest case a pressure pot can be used to good effect.

(a) Mix and meter machines

Mix and meter pumps most commonly use dosing pumps that are connected by a rocker bar or cam. These are available from a large number of suppliers. The mix ratio is set by the individual dosing pump sizes and the position of the rocker bar pivot or cam shape. When rocker bars are used the mix ratio is essentially fixed (or at least is intended to

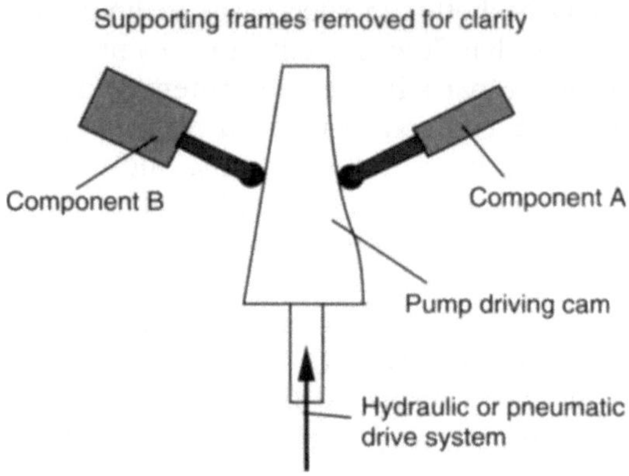

Figure 4.34 Cams can be used to drive RTM mix and meter machines. The driving cam is actuated vertically, driving the pumps for components A and B. Delivery of component B will be constant at a constant drive rate, delivery of component A will vary in line with the cam shape.

be fixed: backlash, seal leakages, lead and lag effects etc. can lead to variations in ratio). If a cam is used the ratio of components can be changed during the stroke (Figure 4.34).

This means that for a polyester moulding the catalyst ratio could be increased towards the end of injection to minimize total cycle time. On the other hand if the ratios are changed this creates a single-shot machine such that the pump sizes would have to be matched to the tool-cavity volume. Other types of metering would include lance cylinders, such as are used in RIM machinery, and gear or peristaltic pumps. In principle injection machinery based on any of these pumping elements would also permit mix ratios to be changed during the resin injection cycle.

The various resin and catalyst streams must be mixed prior to injection. In-line static mixers are most commonly used although dynamic mixers have been used and impingement mixers can be used at very high flow rates. In-line static mixers can give a very good quality of resin mixing, but experience has shown that more static mixing elements may be required than are recommended by the manufacturers. A simple way of checking mix quality is to dispense resin in a line onto a non-porous white surface; this is especially effective if the minor constituent is dyed.

In use, the machines normally pump continuously with the various components recirculated around a loop. When resin delivery is required the two (or more) streams are brought together at the mixer and after injection a flushing cycle may be activated to clear the mixers of resin; a combined air–solvent–air flush is most commonly used. Whether the mixer must be purged after every shot depends on whether the material in the mixer will have started to cure before the next shot is called for.

Polyesters are usually pumped at room temperatures, but heating may be required for epoxies. The simplest way to achieve this is often to encase all the mechanism within a heated cabinet, and additionally use trace heaters on resin delivery hoses, or commercially available heated hoses may be used for the greatest accuracy.

The advantages of mix and meter machines are: instant availability of resin without premixing or precharging; essentially unlimited shot sizes with some variants; high flow rate and pressure capability. The disadvantages are that the pressure drop across static mixing elements can be high so that when the mould is filled the pressure on the tool can suddenly increase to the stall pressure of the pumps. Many modern machines have pressure limiters to overcome this problem. The machines are mechanically more complex than premix machines and need more maintenance. Cleaning out the machines for a change in resin can be a time-consuming business to ensure that all traces of the old resin are removed.

Mix ratio may vary due to seal leakage under operating pressure, air entrainment in resin or catalyst or differences in compressibility between the streams, backlash in linkages, and lead/lag effects. With well designed and maintained machines these problems are minimized, but the fact remains that direct control or monitoring of mix ratio is at best difficult. For aerospace applications it may be essential that a record is kept of the measured mix ratio for each shot. Unless calibrated mass flow meters were fitted to each line this would make the use of mix and meter machines very questionable. Flow meters are available but adequately accurate ones tend to be expensive and premix machines tend to be used in aerospace. Whatever sort of components are being made it is a good practice to conduct a regular check of the mix ratio by drawing off samples of unmixed resin constituents. When this is being done the samples should be taken through a restrictor such that the back pressure is the same as the highest pressure experienced in the mouldings to be made.

Lastly, this sort of machine usually meters by volume and, unless the back pressure capability of the machine is exceeded, a constant fill time will be experienced for a wide range of moulding conditions such as resin viscosity or reinforcement permeability. From production engineering and scheduling viewpoints this is clearly an advantage. However, if for some reason the resistance to flow increases, the increased pressure requirements can lead to premature seal or linkage wear and an increase in machine down time. Changes in pressure cycle can also lead to a reduction in tooling life or problems with moulding quality due to reinforcement movement or tool deflections. If this type of machine is being used without any pressure control a careful eye needs to be kept on any changes in the incoming materials because such changes may be masked by the machine's mode of operation.

(b) Premix machines

The most basic form of a premix machine is a simple pressure pot. Premixed and de-aerated resin is simply poured into the pot and ejected from the pot by air pressure. A cavity in the tool can also be used in the same way, giving an integrated mould and ejector. Alternatively, a vessel of premixed resin can be held in a larger vessel and dispensed through a dip tube. This sort of equipment is cheap, has no moving parts and thus requires little or no maintenance. Very high quality components have been made using the simplest equipment, both in prototyping and production environments. Somewhat more sophisticated systems are commercially available.

The advantages are:

- resin mix ratio can be controlled, monitored and recorded for full traceability (although it must be conceded that manual weighing and mixing of small batches of resins presents many opportunities for error);
- investment costs are low as are maintenance costs;
- simple pressure pots give near total reliability in operation;
- resin type may be changed at will without excessive down time;
- pressure is known at all times and can be controlled so as to prevent damage to tools.

The disadvantages are:

- problems with bulk exotherm may limit shot size;
- available pressure is usually limited, which may limit injection rate, although most RTM applications only require low pressure;
- additional labour is required to premix and charge injectors;
- variations in Vf% of reinforcement or resin viscosity will lead to changes in injection rate, which may cause scheduling difficulties; this problem could be overcome using a flow sensor and feedback control to the pressure: as noted above this ‘problem’ can also be an advantage as any changes in incoming materials or process conditions are immediately obvious.

In addition to the use of gas pressure, various types of pumping element could be used to dispense premixed resins, under pressure or volume control depending on the type of pump used. These sorts of pumped systems will be more complex and expensive and all suffer from the potential problem that resin may cure off in the pumps, necessitating costly repair or maintenance. It would be possible to overcome this problem by using resin supplied in cartridges, driven by an hydraulic or similar cylinder, so that contact between resin and moving parts was eliminated. Lastly, some suppliers of aerospace grades of epoxy now supply premixed resin for RTM; this generally has a very long open time before cure and is ideally suited for use in simple injectors. In addition to the resins, injectors suited to their properties have been designed and used.[14]

Neither of the two main types of RTM injectors can be considered to be ideal for all applications. For each type of machine the limitations could be eased but generally at the price of increased costs, complexity and maintenance requirements.

(c) Interfacing to the tool

Whatever type of resin supply machine is used the resin will generally be supplied down a pipe that must be coupled into the tool. In addition there is often a requirement for a resin overflow line and or vacuum line into the tool. The simplest interface is a tapered injection nozzle that couples into a matching taper on the tool. Most commercial injection machines are fitted with such tapered nozzles and matching injection sprues made from PTFE or similar materials are available from some tooling suppliers for insertion into the tool. If the tool is evacuated a simple matching taper is unlikely to give an adequate vacuum seal and some sealing arrangements are required on the tool or injection nozzle. Figure 4.35 shows a simple version of a nozzle for vacuum applications.

The design of the interface between the tool and injection and any vacuum equipment is in large part a function of the production engineering requirements. Factors such as requirements for the separation of clean and dirty areas, numbers off and cycle time would have to be considered, as would materials issues such as the resin cure time which might dictate whether flushing was required. A few different classes of production engineering requirements are noted below and an attempt made to identify suitable interfaces.

In prototyping, cycle time is often less important than establishing control on the system. If tools are not evacuated conventional tapered seats will be acceptable. For evacuated tools the use of standard pipe fittings screwed into the tool, with translucent nylon pipe to introduce resin and vacuum, has given good results. The advantages are that setting up is simple: flow is visible which is especially useful if several vacuum ports are used which the resin reaches at different times. The flow of resin can easily be controlled by pinching the pipe; pressure can be maintained throughout cure; and some understanding of changes in resin volume during cure can be achieved by monitoring the level of resin in the pipes as cure proceeds. In volume production the use of disposable pipework and fittings would be less likely due to high labour and materials costs, but the added understanding that arises from being able to see the flow out of the tool and during cure is useful in prototyping. If throwaway pipes are used it may be best to carry the pipe all the way to the

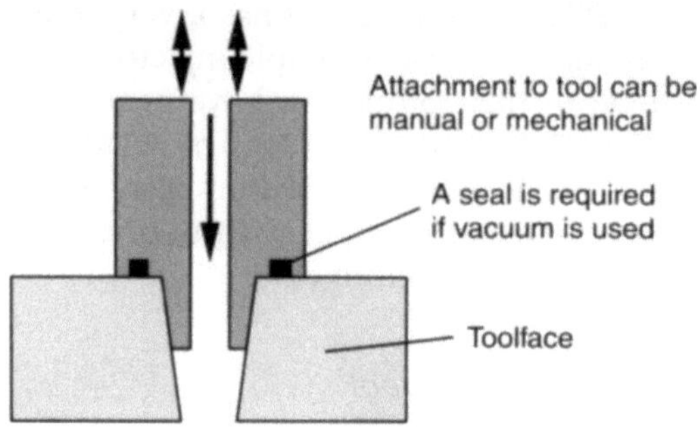

Figure 4.35 A variety of resin supply/tool interface types are possible. The simplest type is shown here.

toolface as this avoids having to extract the core of resin that goes through the toolface if the pipes are terminated at the tool's top surface.

For fairly short cycle times (say 1–5 minutes injection time and 10–20 minutes' cure time) and moderate volumes the normal tapered nozzles would be expected to be used for unevacuated and smaller tools. For evacuated tools the sort of nozzles shown in Figure 4.35 would be used. For the cycle times noted above one injector would be expected to service several tools and demountable injection/vacuum equipment would most probably be used. If the positions of the resin ports are the same on all tools in the system there would be an advantage in mounting the vacuum and injection nozzles on a common frame, which could quickly be secured and demounted from the tools. Various layouts are possible. The tools could be arranged around the injection machine, or the tools could be arranged in a line and serviced from an ejector on an overhead rail etc. At short cycle times it would be difficult to establish a clear separation between clean and dirty areas, although by no means impossible. In general, aerospace parts that require the separation of clean and dirty areas are not made on such short cycle times, so the necessary tool movements can be accommodated. For the cycle times quoted above it is unlikely that flushing injection lines between shots would be necessary if several tools were in operation. It might be prudent to clear resin from any vacuum lines used as this resin will have picked up heat from the tool. This will happen automatically if the vacuum has not decayed away when the vacuum line is decoupled.

If a low level of vacuum is used to assist in the achievement of adequate quality a pneumatic vacuum generator can be used and these are more robust than mechanical pump types. If a high level of vacuum is required a pump must be used and great care taken to avoid contamination of the pump with resin or solvents. In either case an overspill pot of adequate capacity is needed and baffles may be needed to prevent any resin being

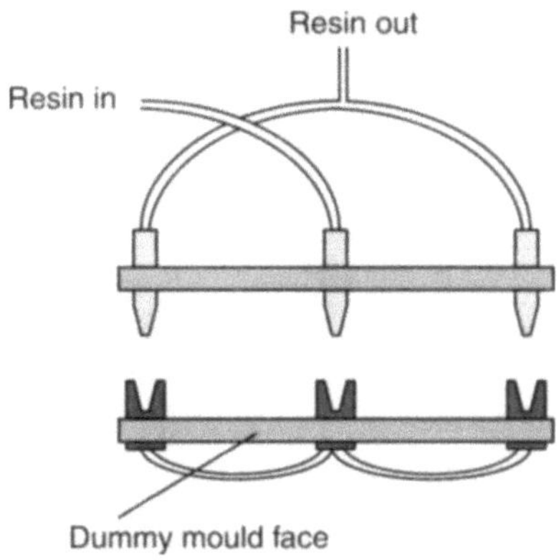

Figure 4.36 If a number of resin flow lines must be cleared at once a dummy mould face can be used to simplify the flushing requirements.

entrained in an air flow into mechanical pumps. The pumps should be turned off or isolated before decoupling the vacuum line from the tool. If a high level of vacuum is required a leakage test should be instituted and ideally the injector should be interlocked to the vacuum system so that injection can only occur when the vacuum parameters are correct. These checks take time so that a high and guaranteed level of vacuum is unlikely to be consistent with short cycle times. Whether or not the system is flushed between shots it will usually be necessary to flush the system at the end of each working period. Lines are first blown free of excess resin, followed by a solvent wash and another air blow to dry the lines. Air used in this way should be dry and oil-free to avoid contamination of injection lines. If a vacuum system is used it would be usual to uncouple the pump and overspill pot before flushing the system. If the injection and resin out/vacuum lines are mounted on a common framework a dummy mould face can be used to ensure that all lines can be flushed at one time (see Figure 4.36).

For very short cycle times it is to be expected that tools will be fixed in place in a mould handling framework or press. In this case the tool is normally hot and permanently fixed injection lines are preferred to minimize handling time. To overcome possible problems with resin curing in the injectors thermally insulated injectors can be used (Figure 4.37) and commercial versions are available.

These insulated injectors could be used on moulds having longer cycle times to reduce labour costs or as part of an automation strategy, but the relative costs of capital and labour would have to be carefully considered. A permanently connected vacuum system would be much harder to achieve, as the overspill resin would have been heated and thus have

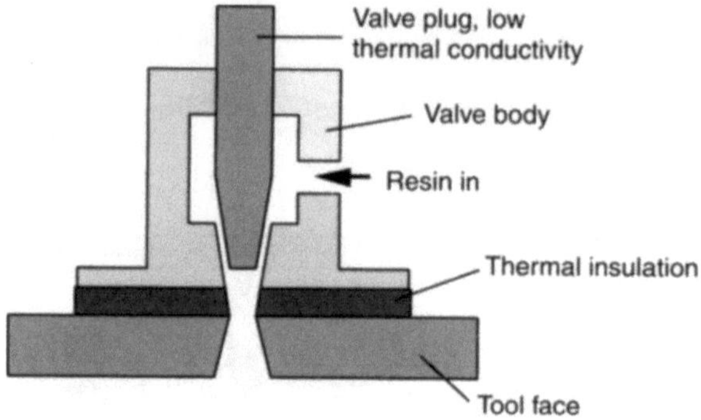

Figure 4.37 If the resin valve must be permanently connected it should be thermally isolated from the tool.

a much higher likelihood of curing off in the pipework. Very fast cycling tools tend not to utilize vacuum, for the reason given above and because of the time required to achieve an acceptable level of vacuum. Very fast cycle tools very seldom make full use of the maximum complexity possible in RTM. In principle, vacuum interfaces can be designed to allow flushing between each shot, but these would be complex in terms of valving arrangements (Figure 4.38).

If it was absolutely necessary to evacuate fast acting tools the best approach would probably be to incorporate an overspill pot into the tool, connected by a serpentine path to the vacuum pump, so that all

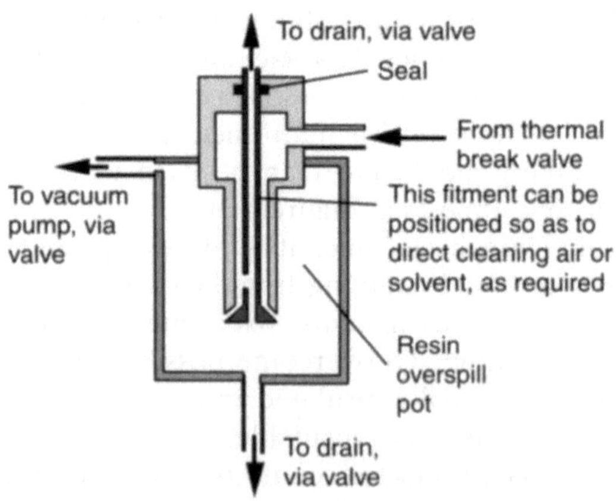

Figure 4.38 While it would be possible to design permanently coupled out-gates in evacuated tools, as shown below, the great complexity makes the use of demountable fittings more common.

the overspill resin was cured off within the tool rather than being carried away to the vacuum system. Resin sensors could also be used to detect the point at which the mould cavity was filled and vent the vacuum lines to atmosphere.

Some tools are fixed in place because of size and weight rather than cycle time. In this case any of the options considered above could be utilized. Any tools that are fixed, for whatever reason, would be expected to lead to problems in an aerospace environment due to the difficulties of separating clean and dirty activities.

In the aerospace environment mould cycle times tend to be much longer and requirements to separate activities usually results in tools being moved around; in addition evacuated tools are commonly used. The sort of simple injectors/vacuum couplings shown in Figure 4.35 can be used. If production volumes are very low, disposable pipework has advantages in that solvent flushing is minimized and the costs of solvent disposal are reduced. The potential for dermatitic reactions to epoxy resins or their solutions makes the use of disposable pipework (which can have the resin

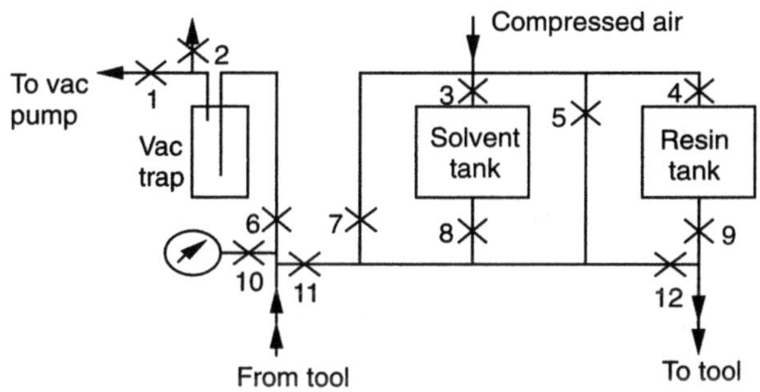

Figure 4.39 If an injection manifold is used to connect the tool to resin, vacuum and flushing lines a large number of valves are needed. The system shown here assumes that compressed air drives the resin pressure. Any valve not noted as open is assumed to be shut.

To evacuate tool open 1, 6, 10.
To inject resin open 4, 9, 1, 6.
To air purge injection pipework open 7, 11, 12.
To solvent purge injection pipework open 3, 8, 11, 12.
To air purge vac trap open 7, 11, 6, 2.
To air purge resin tank open 4, 9.
To solvent purge vac trap open 3, 8, 11, 6, 2.
To solvent purge resin tank open 3, 8, 5, 4, 9.

The number of valves can be reduced by the use of multi-way valves, e.g. 6, 10 and 11 could be combined.

The obvious possibilities for costly errors contribute to the popularity of disposable pipework or systems that cure off any excess resin and avoid solvent flushing.

fully cured within it prior to disposal) a distinct advantage. For quality critical work all the production parameters will need to be checked and recorded. Prior to injection the vacuum level and leakage level should be checked. The vacuum level should be checked at the tool, rather than at the vacuum pump as pressure drops in long vacuum lines can give rise to false readings. Ideally the vacuum gauge should be mounted in the injection port and the vacuum applied to the resin out gate. If the vacuum gauge is to form part of a hard-piped manifold a valve will be required to isolate it during injection.

Figure 4.39 shows a possible set of valve requirements for a hard piped injection manifold.The injection manifold might be coupled up in a tool handling and flushing system using a dummy toolface as shown in Figure 4.36 to ensure that all pipework was cleared; and many other designs are possible. The valving requirements can be quite complex and errors in operation of the valve opening sequence could have serious results for pumps and gauges. Valve gear can be interlocked to prevent errors at an increased cost. For these reasons the use of disposable pipework might be retained to higher production volumes in this environment than would be the case in general industry.

REFERENCES

1. Bergstrom, L. and Alwart, S. (1990) Advantages of mass cast methacrylate tooling for RTM, *Proc. Tooling for Composites*, June, Anaheim: SME.
2. Panzer, D. and Harper, A. (1990) Cost effective composite tooling for the RTM process, *Resin Transfer Molding for the Aerospace Industry*, 6–7 March, Los Angeles: SME.
3. Smith, P. (1992) New developments in composite tooling, *18th International Composites Congress*, British Plastics Federation, Nov., Paper 32.
4. Ridgard, C. (1988) Distortion in composite prepreg mould tools and components with particular reference to low temperature curing and high precision moulding techniques. In F. Saporiti, W. Merati and L. Peroni (eds), *New Generation Materials and Processes*, Milan: Grafiche FBM, 87–100.
5. Peters, D., Bastone, A. and Clark, J. (1990) Aluminium bronze – an alternative tooling material for composites, *45th Annual Conf. Composites Institute*, SPI, Feb., Session 5-C.
6. Roark, R. and Young, W. (1976) *Formulas for Stress and Strain*, 5th edn, McGraw-Hill.
7. See reference 2 (Panzer and Harper, 1990).
8. Nelson, R. H. (1988) Prediction of dimensional changes in composite laminates during cure, *34th International SAMPE Symposium*, May, 2397–2410.
9. *Designing for Diecasting*, London: Fry's Diecastings Ltd.
10. Murphy, J. (ed.) (1991) *New Horizons in Plastics: A Handbook for Design Engineers*, London: WEKA Publishing Group.

11. See reference 2 (Panzer and Harper, 1990).
12. Liu, B., Bickerton, S. and Advani, S. (1996) Modelling and simulation of RTM; gate control, venting and dry spot prediction, *Composites* Part A. 27A, 135–41.
13. Larsen, M. (1990) Injection equipment for resin transfer moulding. *Resin Transfer Moulding for the Aerospace Industry*, 6–7 March, Los Angeles: SME.
14. 3M PR500 epoxy resin datasheet.

5 Production engineering requirements

Production engineering requirements can be very component specific, the interaction of various possible styles of production line with tooling and interface design has been covered in Chapter 4 and will not be repeated. Some general comments can, however, be made here.

5.1 WORKING ENVIRONMENT

Working areas must be well lit, adequately heated and preferably humidity controlled (especially if moisture sensitive cores or other materials are used). Even if rigorous separation of clean and dirty areas is not required some attempt must be made to avoid contamination of reinforcements with debris from deflashing, release agent application or other dirty activities.

The cutting of dry fibre will generate airborne fibre and good levels of extraction and overall ventilation are required. Fibre cutting activities should be monitored to determine the level of released fibres. In my experience the cutting of cloth on knife tools releases little fibre. This fibre seems to remain close to the cutting surface so that sideways or downward extraction would be likely to give better removal than a vertically upward airflow. Whether or not extraction was required would depend on the type and amount of fibre cut and on local regulations as to acceptable levels of airborne fibre. As a general rule, if specific extraction is not required, fibre cutting areas should be regularly cleaned to prevent any build-up of fibres. This cleaning must be done with a suitable vacuum cleaner rather than a brush as brushing will simply make the fibres airborne. The trimming of small amounts of excess fibres from slightly oversized preforms generates a lot more airborne fibre and extraction should usually be provided. The use of a small trimming enclosure with airflow drawn down towards the back of the enclosure seems to give acceptable results.

From an operator comfort point of view it may be convenient to ensure that repetitive tasks such as trimming preforms or mould loading can be done in a sitting position. It is sometimes easier to load smaller tools when they are tilted towards the operator, to avoid back strains from having to work directly above the tool. Lighting in preforming and mould loading areas needs to be very good, especially when carbon fibres are used and accurate placement of preforms is necessary.

5.2 SPECIFIC REQUIREMENTS

5.2.1 Materials storage

Normal clean, cool, dry conditions are adequate for most materials. Some foams may be sensitive to changes in moisture content and controlled humidity is an advantage. Some resins and hardeners are sensitive to moisture absorbed from the air. If these are used they should not be held in large containers as they may degrade over time as the containers are repeatedly opened and shut to draw out small amounts of material. Many resins have to be treated as flammable and normal safety rules apply. The catalysts and accelerators used in polyester mouldings can react explosively if they are brought into contact with each other. Catalysts and accelerators must be rigorously separated in storage. For quality critical work requiring traceability, bonded stores may be required and good control over materials issuing will be critical.

5.2.2 Preform blank preparation and preforming

Good extraction may be required as noted above, especially for preform trimming. Traceability of both materials and process parameters may be required. Safety interlocks will probably be required on preform machinery.

5.2.3 Kit preparation

Assuming that more than one preform is needed to load the tool it is worthwhile to produce a kit of all the preforms that go into each moulding. If kits are produced more or less as required no special attention is required. If a stock of kits is held they should be kept cool and dry to prevent any relaxation from occurring. If the kits are to be stored they should ideally be held in kit trays that provide some support to each preform. Complete kits can be inspected for quality and can be held under a quality seal until use.

5.2.4 Mould loading

As noted earlier, good lighting and a comfortable working environment are essential. If the tool incorporates loose elements there is always a danger of these elements becoming separated from the tool or damaged in some way. Whenever the tool is handled or transported, provision needs to be made to ensure that the loose elements do not become lost: they need to have a proper 'home' on the tool. By definition loose elements should be loose, they should never have to be forced into place and any requirement for force needs to be investigated. Hammers and similar tools should be kept well away from toolfaces at all times. For quality-critical work it is good practice to ensure that the tool serial number is replicated on any loose blocks that are contained within the tool.

5.2.5 Mould transport

Moulds may have to be moved for a variety of reasons. In the simplest case wheeled trolleys may be adequate. In the case of a more formalized production line the use of roller tables can be used to good effect to link the various activities in the moulding cycle.

5.2.6 Resin mixing

If this is required it must be carried out strictly to the manufacturer's safety and handling instructions. Resin mixing as an activity falls somewhere between the normal definitions of clean and dirty. The resin must not be contaminated, so it needs to be kept away from sources of dirt, solvents, moisture etc. On the other hand, as far as the dry reinforcements are concerned, the resin is a potential source of contamination. Ideally, a separate resin mixing area should be used, taking care not to disrupt the production flow by siting resin mixing at any distance from resin injection. If resins are mixed prior to use the control over the mixed resin's thermal history after mixing must be rigorous to avoid any element of pre-cure prior to injection. Standards are required for how long resin can be held mixed prior to use and must be strictly adhered to. Premixed resins will also normally require a vacuum degassing stage prior to injection to remove any entrapped air. As before, standards are required and should be adhered to.

5.2.7 Resin injection and curing

Resin injection is widely reported as a clean process that avoids contact between resins and the world outside the RTM tool. In practice I have seen many RTM facilities and most of these show evidence of contamination with resin and solvents. When used carelessly the claim of RTM to be a clean process is insupportable, even when used with care some small spillage is commonplace unless tools are close-coupled to injectors. For this reason it is better to keep dry fibres and preforms well away from areas where resin is being injected. Any resin spillage that does occur should be dealt with immediately, to keep the workplace clean, to minimize styrene levels if polyester resins are used and to minimize dermatitic reactions with epoxy resins.

5.2.8 Demoulding/deflashing/mould cleaning and preparation

These activities inevitably generate debris, such as resin flash, that must not find its way into future mouldings. Mould cleaning materials and release agents would also be severe contaminants of dry preforms. Dry fibres and preforms must be kept away from areas where these activities are taking place. It bears endless repeating that all tools, equipment, methods and materials used in demoulding, mould cleaning and release agent application must have been approved by the tooling designers as appropriate; taking into account the tooling material and mould design. It takes only a moment's carelessness or inattention to scratch or damage a sealing surface or tool surface with inappropriate tools. No tools or other materials such as solvents that have not been approved should come anywhere near the mould surface. This may mean banning such tools and materials completely in the area reserved for these activities. If this stricture seems unnecessarily paranoid readers are invited to permit the use of any means the shop floor workers desire in demoulding and mould cleaning, provided that the tools have no value to them.

5.2.9 Shop floor layout

The details in this area depend on very many factors, from the size and weight of the tools, the heating strategy used, quality requirements etc. The approach taken depends on production volume and cycle time and has been outlined in section 4.3.11.

Figure 5.1 shows a possible layout for an injection shop that is operating on a fairly short cycle time for general industrial products. Figure 5.2 shows a possible layout for an injection shop that is operating in an aerospace environment. There are very major differences between plants that must operate with a formal separation between clean and dirty areas

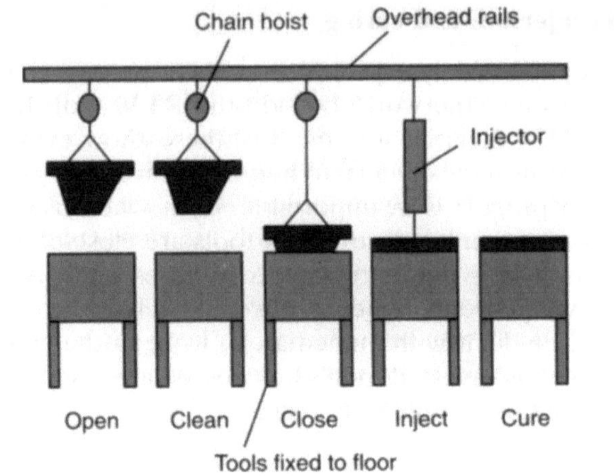

Figure 5.1 A simple layout for a set of production tools is shown here. In this case these is no separation between clean and dirty areas, separate provision is required for mould clamping, and if heating is required each tool may require integral heating.

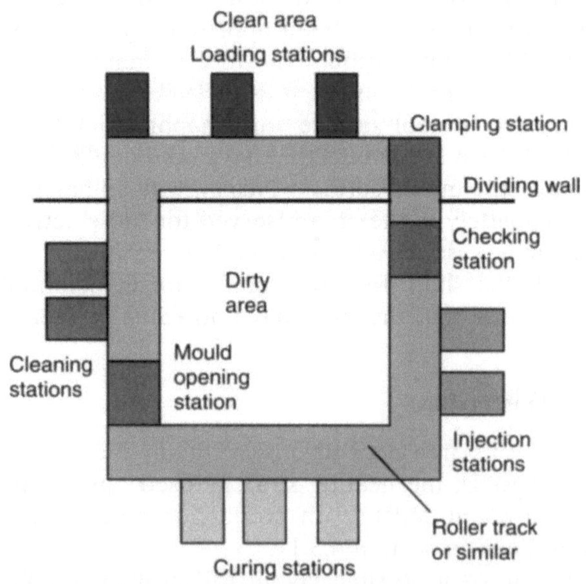

Figure 5.2 In this layout clean and dirty areas are separated, as would generally be required in aerospace. The number of stations for loading, injection, curing etc. depends on the number of tools in the circuit and the relative time taken by each activity. Such a layout is not likely to be economic at low volume.

and those that can use good housekeeping practices to avoid cross-contamination. Formal separation might be expected to increase the requirements for floor area, increase capital and labour costs for tool handling, perhaps lead to a duplication of some facilities and potentially lead to a less flexible production environment in the injection shop (i.e. less easy to reconfigure or expand). In addition, formal separation might influence some areas of the tooling design. In general terms these factors can be expected to increase some areas of the total product cost. For this reason the decision as to whether formal separation is required must be made at the start of the product development process, and not left to the point that detailed plans for production facilities are being made. In the worst case, starting up a production facility assuming no separation and then having to impose it can lead to much more expense than starting out with this as a requirement.

Obviously, while the injection shop is central to an RTM production line the other elements of reinforcement preparation and moulding finishing are just as important to the overall economics. These can be dealt with in much the same way as they would be for any composites production line. The general rules are to co-locate facilities to avoid long transport times for materials or mouldings and to ensure that the movements of materials, tools, mouldings etc. take place in a logical manner around the cycle. In common with most composites manufacturing processes an RTM facility can be set up very quickly once the tools are available. If timescales are tight this is often seen as an advantage. However, an RTM line that is rushed into operation without a lot of thought as to the proper production engineering is unlikely to run efficiently or be easy to control. In this case it is more than likely that the line will eventually have to be shut down and re-engineered, inevitably at a higher cost than that of doing the job properly from the outset.

6 Component design for RTM

As in previous chapters the baseline RTM process that will be considered here is the use of essentially rigid tooling which generates a fixed dimension mould cavity. The use of non-rigid tooling is discussed in the next chapter. It is also assumed that high-performance components are required, whatever the materials of construction. For this reason the comparison is made below between autoclave and RTM design practices, rather than for example between contact moulding and RTM.

Autoclave moulding uses preimpregnated materials which already have a high added value, leading to a reasonable desire to use them as weight efficiently as possible. The design approach is based largely on composites theory, tempered by experience and some empirical rules. In essence, each ply is treated as an independent unit of the overall construction and placed in the most favourable position and orientation with respect to the tool and the plies that surround it. This leads to layup of one ply at a time, with regular pressure cycles (in some cases every three or four plies) to remove most of the bulk factor in the prepreg (bulk factor is the thickness of the as-received ply divided by the cured thickness). Debulking is in part a matter of resin flow so that the process is not instantaneous and debulking cycle times are often many minutes rather than the seconds used in preforming dry reinforcements. These features often lead to long layup times even for small parts.

Having produced the layup, consolidation is achieved by the use of air pressure in an autoclave or vacuum bag. Even with many cycles of debulking some further consolidation will be required during the cure cycle. As consolidation takes place various defects in the component geometry can be generated. Wrinkles may be formed on external radii and bridging of fibres or bagging materials across internal radii is common (Figure 6.1).

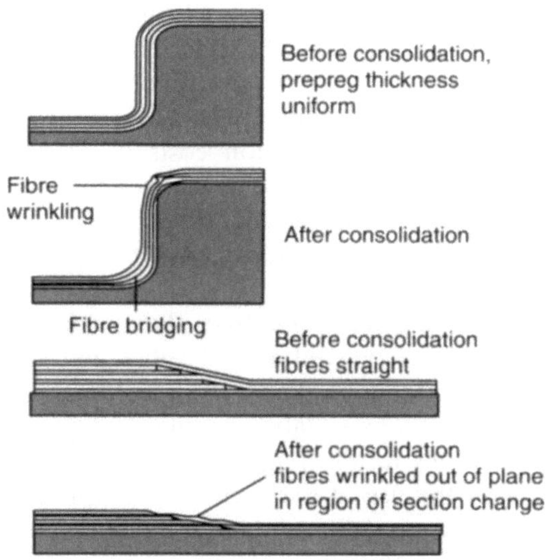

Figure 6.1 Various defects can arise during consolidation of autoclave moulded components. These can reduce component performance.

These defects can be controlled by good operating practices and by some modifications to the layup, at additional product cost. Overall these effects limit the complexity that is available for autoclave moulded components of the highest quality, so that complex parts may comprise several sub-components. Apparently small increases in complexity can have disproportionate effects on the costs of autoclave moulded components, in terms of high scrap and rework rates or higher labour or tooling investments to eliminate the areas of process risk. In part, these relationships drive the decisions as to whether to use parts integration or make separate parts and attach them together. Other influences on this decision might be concerned with the ease of increasing throughput in the two cases etc. The same questions apply to RTM. It is by no means always a good practice to use the maximum possible level of parts integration, but the driving forces are different for RTM leading to different solutions in some cases.

It can be seen that the design of components for autoclave production is constrained by the nature of the materials and process used. Despite this, because the process is primarily used for the highest quality components, there is a tendency to assume that autoclave design practices are equivalent to best design practices for high-performance components. This is not necessarily the case, especially when production costs and scrap and rework rates are taken into account in deciding whether a component is considered to be of high performance.

In contrast to autoclave moulding which emerged from the materials science community and found use primarily in aerospace, RTM emerged from the general moulding community and initially found use in that product area. The main driving force behind the RTM developments has been to reduce the production costs of component manufacture. Other development areas have been to broaden the geometrical possibilities and give greater scope for parts integration and to utilize a wider range of reinforcement styles, but the primary focus has been on cost.

Some cost reductions may come from the avoidance of preimpregnated materials with their associated added value and storage and handling costs. Some cost reductions come from the increased productivity of mould tools or reductions in labour contents. Some cost reductions can come from reductions in finishing operations, improved quality or parts integration. Lastly the use of the preforming approach to building up the layup prior to injection can have a major effect on costs. Set against these areas of cost reduction there are, of course, areas of cost increase; notably higher costs for mould tools.

While it is perfectly possible to design high performance components for manufacture by RTM to the design standards used in autoclave moulding, these practices can lead to some problems. Many of these have been noted earlier, such as sensitivity to local changes in Vf% and a preference for surface-mounted ply drops in high Vf% components. The use of geometrical features, such as large bend radii and the general avoidance of undercuts carried over from autoclave moulding, must also be questioned. Lastly, the use of the single ply as the basic unit of the layup is very restrictive compared to the use of blocks of plies converted into handleable preforms (or the use of other preform types such as net-shape braided reinforcements). In order to take full advantage of the cost-reduction potential of RTM some changes have to be made in the design approaches used.

A few statements might be made to illustrate the basic approach to the design of cost-effective high-performance structures via RTM. Other advantages of RTM such as well-toleranced dimensions may lead to cost savings; via a reduction in finishing operations or pre-assembly shimming; but do not so directly influence the design process.

- Ply-by-ply design is structurally the most efficient, although its theoretical efficiency can be severely degraded by moulding defects associated with layup and debulking.[1]
- Ply-by-ply design tends to lead to high labour costs. Problems with moulded quality can also increase disproportionately as complexity increases, the steps taken to overcome these problems can cause great increases in labour contents.

- Layup of blocks of plies reduces layup time in proportion to the number of plies in a block. Manual layup of blocks of prepreg plies is very difficult because of the need to ensure that out of plane shear occurs between the plies without wrinkles being formed. It is possible to form blocks of prepreg mechanically in some cases, but this approach has not been extensively developed and the use of manually emplaced single plies is the norm.
- Preforming ply blocks into the required shape is a much more rapid process than manual layup of ply blocks, and thus further reduces labour costs.
- Complex structures can be formed from an assembly of preformed ply blocks and moulded to reliable quality levels.
- The possibility for additional complexity and reliable quality reduces the costs of parts integration and makes this a more attractive option.

Many of these approaches could also be used with prepreg, and similar approaches are used with resin film infusion; an autoclave based process having some of the features of RTM. In the case of autoclave moulding the problems of debulking inducing wrinkles and bridging would still be present, and could be worse if ply blocks are used.

A move away from the use of the single ply as the basic unit of reinforcement has some obvious potential problem areas. The ends of the ply blocks would be expected to be stress raisers. Resin rich zones can be expected at the ends of ply blocks; curing stresses can lead to cracking of these resin rich zones and these may propagate under applied mechanical or thermal stresses. Proper positioning of ply blocks so that they do not end in peak stress areas, adjustment of the local part or tool geometry and the selection of tough resins can limit these problems, as can the use of surfacing materials where ply-block drop-off points are surface mounted. It should be noted that very many components are stiffness limited and operate at very low strains and while ply block ends will act as stress raisers it is far from axiomatic that they raise stresses to the point that problems would occur. Practical experience with complex components made from ply block preforms has shown that resin cracking can occur in resin rich zones without necessarily leading to structural degradation within the design envelope; even after thousands of freeze-thaw cycles.

When using ply-by-ply design it is possible, in principle, to ensure that each ply is optimally positioned and oriented such that each ply carries its full share of the imposed loads. In practice some deviations from the optimum are usually required for reasons of layup difficulty or cost. The approach might be characterized as first designing the best structure and then modifying it for the purposes of achieving acceptable reliability in manufacturing. This is obviously a caricature of the way that first-rate designers would design a part for autoclave production, but I have seen very many instances of parts which have had to be extensively modified for reasons of manufacturability and hence costs. By contrast a designer would have to be very lucky to find that an assembly of ply block preforms was also an optimum structure. The RTM design approach might be characterized as designing the best structure for manufacturing and then modifying it to improve the structural performance. This is also a caricature of the way that RTM parts are designed, but equally contains more than a grain of truth.

In general the design approach for the RTM production of complex components via an assembly of preforms route seeks to trade off the perceived – but not always realized – structural efficiency of the single ply approach for efficiency in manufacturing. Great care is clearly needed in making these trade-offs as both structural and cost targets must be met. Even in the aerospace field there is now a much greater emphasis on manufacturing costs than was historically the case. The maximization of manufacturing efficiency is therefore of great importance, and designing to make the best use of RTM technology – rather than following custom and practice solutions – will be of increasing importance. Outside aerospace, costs are usually of paramount importance and making the fullest use of the flexibility inherent in the RTM process seems to be one of the very few ways in which the use of composite materials can be greatly expanded.

RTM components can be of more complex geometry than those made by most other routes. If the components are carrying appreciable loads detailed stress analysis will be needed. 3D FEM is often a requirement for these complex parts and may be an absolute requirement if thermally induced stresses must be accounted for. FEM is primarily a method of checking the design in later stages of the overall design cycle. Simpler tools are needed in the early stages of design when the size and shape of possible preform blocks are being considered. One method that can be used is to visualize each preform block as a separate, thin, constant thickness component. The total structure is then visualized as comprising an adhesively bonded assembly of these components. The usual rules for adhesively bonded structures, such as the avoidance of peel or cleavage stresses, can be used to give a view on the likely adequacy of the overall structure (see Figure 6.2).

This approach is no substitute for a thorough stress analysis but can be of great assistance in the conceptual stages of design. It is not absolutely necessary that all the preforms be in effect surface bonded to each other – final changes to the preform geometry to improve fibre continuity or for other reasons can be made at the mould loading stage (see Figure 6.3). Some conservatism in highly loaded regions or very complex regions of RTM components is generally advisable. If this is

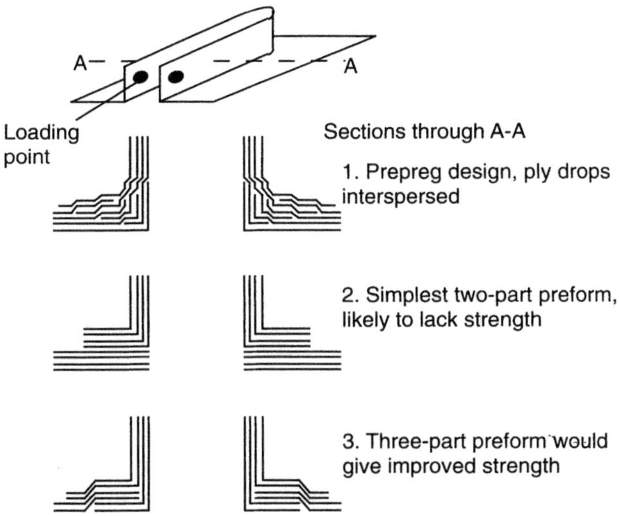

Figure 6.2 Designs in prepreg usually intersperse plies, adding to costs. The lowest cost preforms designs may have to be modified to give adequate strength. Note, prepreg design would require substantial radii at corners, a minimum of about 5 mm would be used here. The preforms could be produced with much tighter radii if this had advantages.

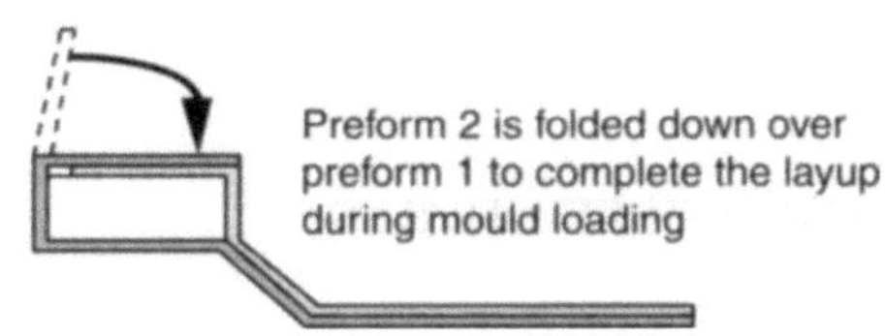

Figure 6.3 Some final modifications to preforms can be made during mould loading, for example to improve fibre continuity.

allied to a good understanding of the capabilities of preforming the objectives of achieving both reliable performance and acceptable costs should be met. Even though RTM is capable of giving high levels of parts integration, this ability should not be pursued for its own sake, but only when it leads to low cost coupled with reliable performance. If a post-moulding bonding operation results in lower overall costs, improved quality or greater ease in accommodating rate changes then this would be a better solution than the single integrated component.

It has been noted that complex autoclave moulded parts can experience quality problems leading to high scrap and rework rates. It is worth while at this point to spend a few moments considering scrap and rework rates in RTM. Parts can be rejected (and then either scrapped or reworked) for a variety of reasons. Neglecting machining-induced damage and rejections for moulding cycles outside preset process limits, the most common causes of rejection for autoclave moulded parts are probably dimensional errors, especially on the thickness. Other causes of rejection might be surface irregularities or roughness, fibre misalignment, wrinkling or bridging, poor bonding of honeycomb cores or crushing of cores and resin starvation. The dimensional errors, including surface irregularities and bridging, will be largely absent in rigid tool RTM mouldings, removing some major quality issues. Up to some point all the non-dimensional errors are acceptable and the point at which they become unacceptable is defined by the Acceptance Criteria used by Quality Inspectors. RTM components are by no means immune from defects and may show defects such as incomplete mould filling that are impossible in an autoclave. On the other hand the general geometry and surface finish of RTM parts made on good tooling will generally be better than that of autoclave moulded parts. This makes even very minor blemishes more apparent on RTM parts. If cosmetic finish is critical these blemishes may be important, for structural parts the Acceptance Criteria should be based on engineering grounds rather than cosmetic appearance. With good tooling, good process control and a well-trained workforce reject rates in RTM moulding lines can be very low compared to autoclave moulding lines, to the point that rejection is a rarity rather than a daily experience.

6.1 SPECIFIC DESIGN FEATURES

Many specific design features have tooling approaches associated with them. Where these have not already been covered in Chapter 4 the tooling features will also be discussed here.

6.1.1 Scale of components

The largest RTM components can be many square metres in area. These tend to be of low Vf% if fabricated in rigid tools. For higher Vf% and large areas, flexible or semiflexible moulds would often be used. These are covered in Chapter 7. The majority of complex or high Vf% RTM mouldings are smaller with resin injection lengths up to around 1 m. For larger mouldings multipoint injection might be used. Alternatively moulds can be injected slightly open to increase the reinforcement permeability, and closed fully at the end of injection. Compliant seals would then be required and mouldings could not be of the very highest complexity. The thickness of components known to me varies from under 1 mm to 80 mm, this range in no way indicates the limits of the process, merely one person's experience. Rapid changes in section thickness can also be accommodated in RTM mouldings.

6.1.2 Corner radii in RTM components

Corner radii achievable in RTM processing are entirely a function of the tooling as bridging of corners is not a problem. Potter and Robertson (1987)[2] quotes an internal bend radius of 1 mm on a 2 mm thick wall thickness. This is a bend radius of 0.5 times the thickness as opposed to the five times the thickness usually used as a minimum for autoclave moulding. The reasons for high-bend radii in autoclave moulding are usually quoted as being a combination of the potential for bridging, the minimization of thermally induced out of plane stresses and resistance to shear failure from imposed loads carried around the corner. I have seen many RTM components in which corner radii were substantially below five times the laminate thickness, without seeing any evidence of thermally induced interlaminar cracking. Speculation as to why this should be so in the absence of detailed modelling is futile, but my personal opinion is that perhaps the through thickness strength of autoclave moulded corners is degraded by a reduction in consolidation pressure, even when bridging is not immediately apparent. Whatever the reasons, the adoption of a standard radius of five times the thickness is unduly conservative, unless loads are carried around the radius, when detailed modelling of thermal and imposed loads would be needed.

6.1.3 Net shape moulding

RTM offers the possibility for net shape moulding to minimize post-moulding operations. See section 4.3.6.

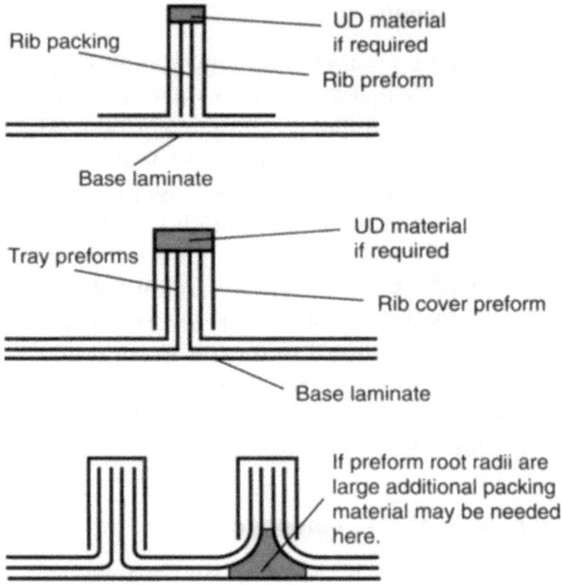

Figure 6.4 Various options are available for the generation of rib structures in RTM. Hollow or foam filled ribs are also possible.

6.1.4 Ribbed structures

Both simple and complex, intersecting, rib structures can be moulded via RTM. The use of ribs complicates both preform manufacture and the determination of correct in and out gate positions. These difficulties can be overcome to generate structurally sound complex structures. Two forms of ribs can be utilized, solid ribs and ribs with foam cores.

Figure 6.4 shows some of the options for the manufacture of solid ribs, via a preforming approach. The outer form of the rib can be constructed from preforms with internal packers of suitable reinforcement, these can be conveniently die cut to the appropriate size. Some care has to be taken in this as some thick, die-cut layups have been seen to emerge from the die with a slight taper in the cut direction. The alternative is to use preform trays with external cover preforms to protect the edge of the rib. The root radius at the base of the rib is an area requiring some comment. Very small radii can be made via RTM, which avoids potential problems with internal resin rich zones where preforms come together. If radii must be larger, some form of packer will be required to avoid the formation of large resin rich zones, and to ensure that preforms do not deform out of plane into unsupported regions (Figure 6.5).

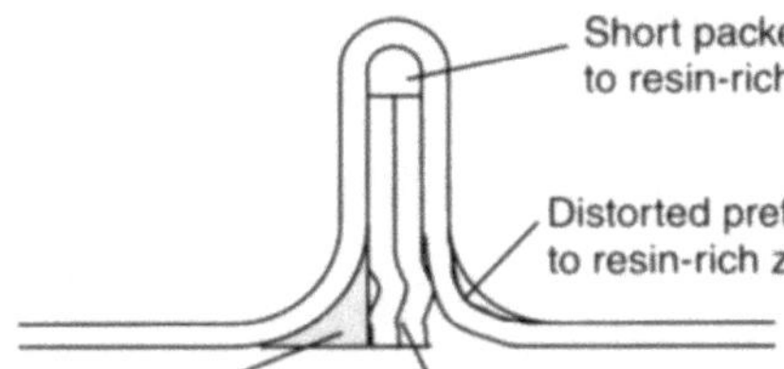

Figure 6.5 Various faults can be generated during the mould loading stage when preforms are used.

Equally, if ribs intersect, care must be taken to ensure that there is adequate fibre continuity around the rib intersection. Problems can be experienced in loading rib preforms into the tool, especially if draft angles are low and the target volume fraction is high. This can show up as wrinkled fibres in the rib packers, or an inability to 'bottom out' the packers. Either effect will lead to resin rich zones and possible quality problems. The use of a reduced Vf% in the rib can be a help in avoiding these problems, as can the use of felt materials within the rib layup. A lower Vf (and hence higher permeability) can also be of assistance in ensuring good wet-out of ribs. Ensuring good impregnation of rib structures without air entrapment can be a problem. Figure 4.33 shows how some gating options are almost certain to lead to air entrapment. For complex rib structures, the best place to inject resin may be at the top of the ribs, where the tendency towards a lower Vf% can lead to the ribs acting as a gating network. Alternatively, resin out ports could be sited at the areas of potential air entrapment. In any case, it is likely that very complex rib structures will require the use of evacuated tooling.

The use of foam-cored ribs reduces potential problems with resin richness at the root or if higher draft angles are used. To keep point-to-point permeability constant, the tolerances on the foam must be kept very close to ensure a constant Vf%. Just as for solid ribs, great attention must be paid to the avoidance of air entrapment. Because the flow front must split completely around the foam core, this can be a greater problem than for solid ribs. In addition, any differential pressure across the foam during injection can cause the foam to shift in position (see next section).

6.1.5 Sandwich panels

The conventional core for sandwich panels in autoclave moulding is honeycomb. Untreated honeycomb is not usable in RTM, because of the

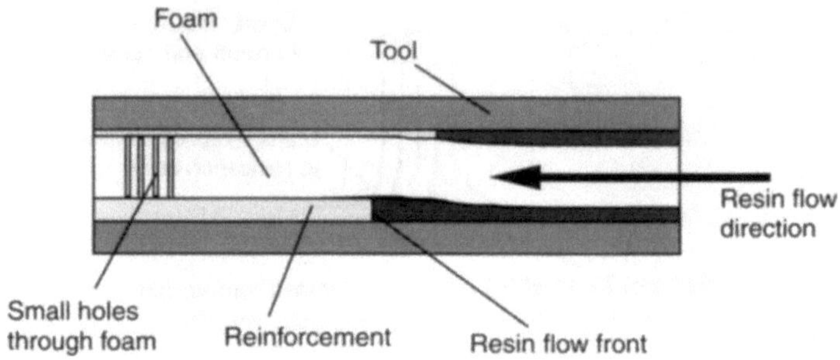

Figure 6.6 If foam-cored sandwich panels are being made the foam can be displaced by uneven flow, resulting in different skin thicknesses and disrupting an even flow front. Displacement happens only if the reinforcement is under a low compaction pressure relative to the injection pressure. The problem can be reduced by piercing the foam to equalize resin pressure.

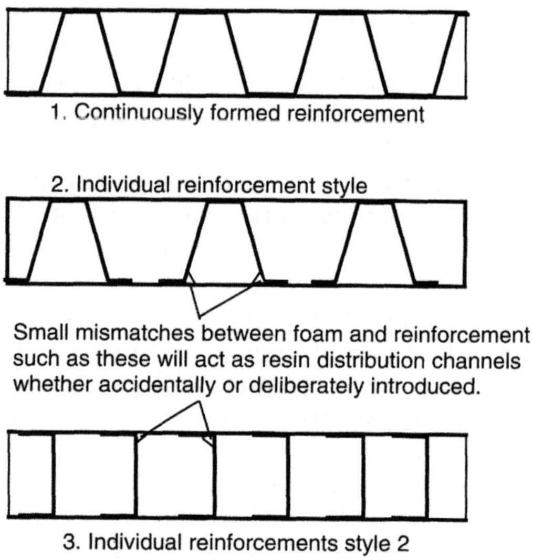

Figure 6.7 Various styles of internally reinforced foam cored panels are possible.

open cells. I have made panels in which a film has been bonded to the honeycomb with film adhesive prior to resin injection. These panels gave similar strength values to panels made in the autoclave even after full moisture conditioning, but the necessity for additional processes and materials makes the use of foam cores more common. When processing foam cored sandwich panels flow of resin is taking place on both surfaces. If the flow front on one side passes the flow front on the other side a force is generated that pushes the foam towards one side of the tool (see Figure 6.6).

This will have the effect of reducing the permeability on that side and thus provide positive reinforcement to the effect. This is less likely to be a problem if the fibre compaction pressure is high compared to the injection pressure and can be relieved by providing passages through the foam which can equilibrate the pressure. While foam is being increasingly accepted as a structurally sound material it may sometimes be necessary to incorporate internal reinforcement webs. Figure 6.7 shows various styles of these.

6.1.6 Handling undercuts

Undercuts are commonly features of RTM mouldings, particularly when a great deal of parts integration is attempted. Potter and Robertson (1987)[3] shows such a component. The handling of undercuts is essentially a mould and preform design task. In some components undercuts are an inevitable part of the geometry, in others they can be removed by minor geometry changes (Figure 6.8) or the use of a foam core to fill in the undercut (Figure 4.11).

In so far as undercuts always tend to increase tooling costs and labour costs it is well worth while to consider alternatives. However, RTM is well able to cope with undercuts and similar complex geometries and it is unnecessarily restrictive to attempt to avoid all undercuts.

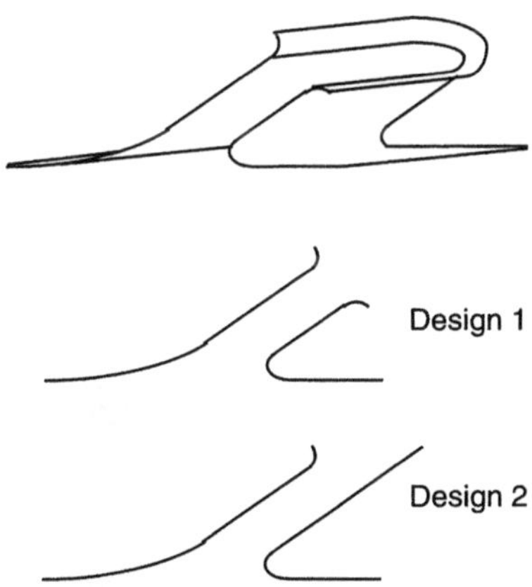

Figure 6.8 In some cases changes to the part geometry can be used to simplify tooling and reduce the number of tooling blocks.

6.1.7 Loading points

RTM mouldings can be treated in the same ways as other mouldings with respect to the introduction of loads and connections to other components. The ability of RTM to generate accurately flat flanges can be of assistance in both adhesively bonded and bolted connections. Moulded in steps and ledges can also simplify post-moulding and assembly operations. In addition alternative options are opened up by the possibilities for complex components inherent in RTM. For example, Morgan (1989)[4] describes the use of moulded in loops made of unidirectional CFRP to form a load carrying bearing attachment (see Figure 6.9).

Potter (1985)[5] describes the use of a moulded in castellated ring to carry the webbing set in a protective helmet (see Figure 6.10). Sealing surfaces can also be considered here; 'O' ring groove tolerances can be achieved via RTM,[6] and other sealing seat geometries such as chamfers can also be achieved.

6.1.8 Inserts

Inserts can be composite or non-composite; moulded in holes can be considered as equivalent to inserts as the techniques required are the same. Potter (1986)[7] describes how a composite insert was used to produce a fine tolerance bore for an interference fitted metal item. It would have been possible to mould the bore directly into the component, but the separate manufacture of these items allowed process risk to be

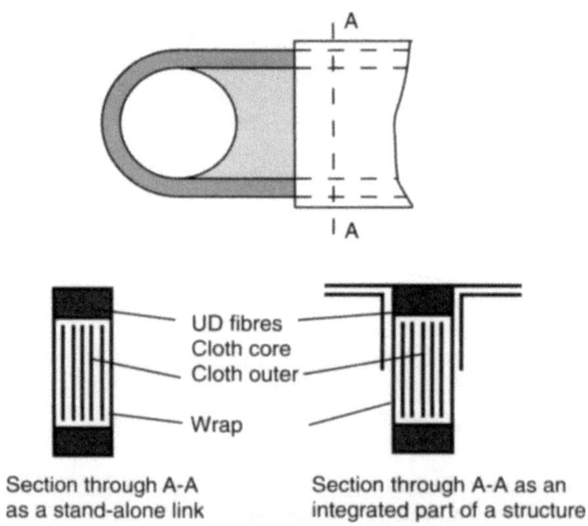

Figure 6.9 Designs such as that shown below have been used to carry loads into RTM structures and diffuse those loads into the main structural elements.

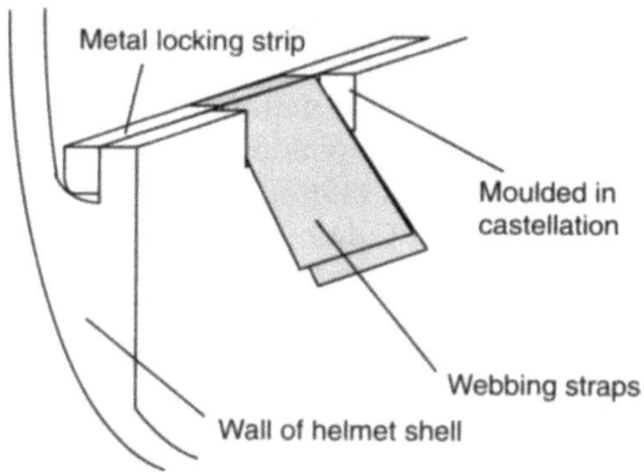

Figure 6.10 Complex, moulded-in castellations have been used to carry loads into RTM moulded helmet shells.

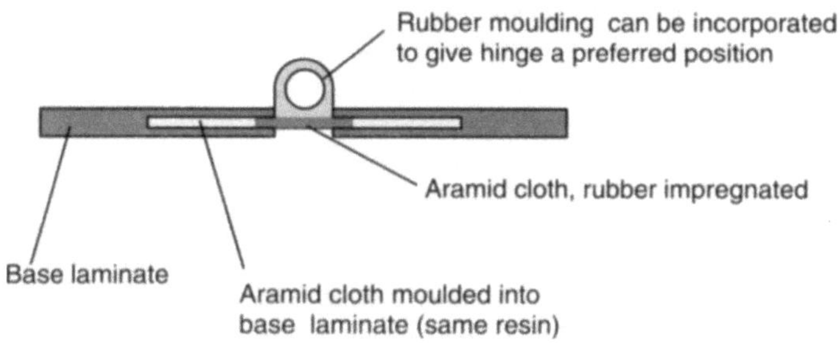

Figure 6.11 Inserts of rubberized aramid cloths have been used to create moulded-in hinges in RTM mouldings.

minimized as the most quality critical parts of the finished moulding could be produced off-line and carefully inspected prior to moulding the higher added value component. In this case prudent considerations of process risk led to a decision not to utilize the full capability of RTM for parts integration. In general, most inserts are metallic and severe problems can arise due to mismatches between the metallic and composite thermal expansion coefficients. These problems are less common for glass/polyester mouldings, but CF/epoxy components can give rise to substantial mismatch stresses with metal inserts. Titanium is a reasonable match to quasi-isotropic CFRP, but not to UD material. Steel and aluminium are increasingly poor matches to CFRP, but steel is a fair match to GRP. Section 4.3.9 discusses some factors related to tooling for inserts.

Holes and internal ducts have been made using melt-out cores, removable mandrels of PTFE or rubber, or by incorporating pre-made ducts. Acoustic panels have been made with a multiplicity of holes by a variety of methods including melt-out pins and both rigid or flexible (injection moulded) pin-boards. Morgan (1989)[8] describes such a perforated panel for a production component. Other materials that have been inserted into RTM mouldings include strips of reinforced elastomer and partially rubber impregnated aramid cloths. These have been used to form moulded-in hinges between two rigid sections of the moulding (Figure 6.11).

Such integral hinges have been produced as prototypes, but it is not known whether such designs have been used commercially.

REFERENCES

1. Hart-Smith, L. (1988) Designing with advanced fibrous composites, *Proc. Australian Bicentennial International Congress in Mechanical Engineering.*
2. Potter, K. D. and Robertson, F. C. (1987) Bismaleimide formulations for resin transfer moulding, *32nd International SAMPE Symposium*, April.
3. Ibid.
4. Morgan, D. (1989) Design of an aero-engine thrust reverser blocker door, *Proc. 34th International SAMPE Symposium*, 2358–64.
5. Potter, K. (1985) Advances in the resin injection process for complex components and their interaction with the design process, *Proc. 1st European Conference on Composite Materials*, 24–27 Sept, 681–6.
6. Potter, K. (1985) Advances in the resin injection process for the reliable production of complex structural components, *Proc. 6th International European Chapter Conference of the Society for the Advancement of Material and Process Engineering*, 247–54.
7. Potter, K. (1986) The development of a GRP casing component and its manufacture by the resin injection moulding process, *Proc. 15th Reinforced Plastics Congress*, British Plastics Federation, 123–5.
8. See reference 4 (Morgan, 1989).

Flexible tool RTM | 7

To a large extent, the use of flexible tool RTM relates to the manufacture of larger components, where either the costs or clamping pressures required for rigid tool RTM are excessive. Equally, for smaller components required in small quantities the use of lower cost flexible tooling may be an advantage. The emphasis in this section will be primarily on the use of flexible tool RTM to manufacture large components. Basic RTM theory still holds for these components, but some additional considerations must also be taken into account. The point at which components become 'large' is not a simple matter of choosing some value of m^2 above which the technology changes as, if production volumes are high, rigid tooling may still be preferred for 'large' components. Having said this, very large components are seldom required in high volumes and flexible tooling will often be the only cost-effective solution.

If the tooling is flexible the consolidation pressure to achieve the required Vf% will generally be limited to vacuum pressure, although autoclaves or pressure boxes could be used to provide consolidation pressures in a hybrid process. Equally, large components need large sheets of reinforcement and large volumes of resin must be handled. These factors then have a strong influence on materials selection, materials handling, preforming, resin delivery machinery and tool design. Some of these influences are considered below.

7.1 MATERIALS

Large area fabrication tools would be expected to be difficult to heat to the temperatures normally associated with epoxy curing, although the lower temperature variants may be more easily handled. Other factors in matrix choice will be minimization of viscosity and control of cure time. While many techniques are possible to increase the speed of injection, it is likely that injection times will be substantial and a balance must be struck between resins with a short cure time and any additional complexity in tooling or process that must be introduced to cope with relatively short injection times. For these reasons polyester and vinyl ester resins are perhaps the most likely candidate matrices for large area work. The use of such styrene containing resins places some limitations on maximum vacuum levels. For example, the vapour pressure of styrene is 13 mbar at 30 °C and 53 mbar at 60 °C. To avoid potential problems, vacuum levels below 900 mbar are likely to be required, i.e. effective consolidation pressures of 0.9 bar. Figure 2.5 shows the effect of such a limitation on the achievable Vf% for various reinforcements. If the components are both large and highly loaded, one of the elements in reinforcement selection will be to choose reinforcement forms with a high natural packing density. Equally, reinforcements with a high permeability for a given Vf% will be preferred and the techniques noted in section 2.1 will be of value. In the simplest version of flexible tool RTM a thin vacuum film might be used to form one toolface and resin simply drawn into the tool cavity by the action of the vacuum in the tool. In this case the effective applied resin pressure is equivalent to the reinforcement consolidation pressure. Increasing the applied vacuum level both increases the driving force and reduces the permeability (via increasing the Vf%)

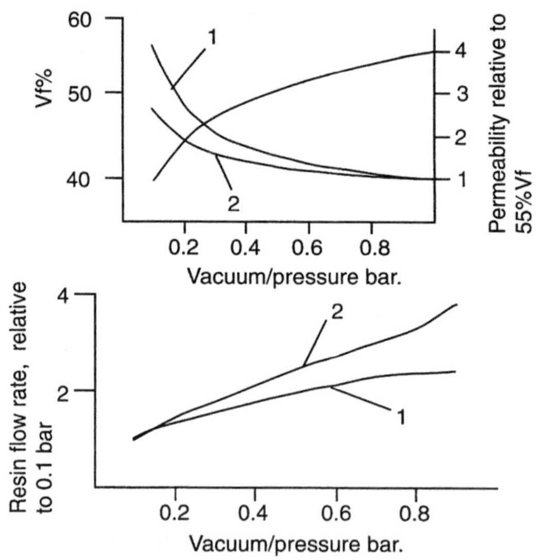

Figure 7.1 For vacuum only RTM in flexible bags the injection rate is not a simple function of pressure as the applied pressure also consolidates the reinforcement. The upper graph shows the effect of pressure on Vf% and hence permeability for two reinforcements. The lower graph shows the combined effect of pressure and permeability. For the material properties shown here a ninefold increase in resin driving pressure produces only a fourfold increase in flow rate.

such that large increases in applied resin pressure result in only small increases in resin flow rate (see Figure 7.1). If the alternative options for increasing resin flow are not used the reinforcement design will be a major element in the overall process design.

7.2 MATERIALS HANDLING

For very large mouldings the use of the preforming techniques discussed earlier will be very difficult. This is true both from the viewpoint of producing preform tools and of the handling and manipulation of large preforms. In some cases direct manual layup into the mould tool may be acceptable. If mouldings are complex a sequential preforming approach can be used, where a tool vacuum bag can form the material (see Figure 7.2).

If very accurate preforms are needed which require trimming it may not be possible to make these sequentially, but they can still be made within the mould tool in the same way. It is possible to trim in the tool by sliding a steel shim under the edge of the preform to be trimmed, but great care has to be taken to avoid damaging the tool and to ensure that all the trimmings are removed from the tool cavity. The use of thermally activated binders may be impossible in this case, but water-carried emulsion binders have been used with vacuum bag preforming. Of course, if

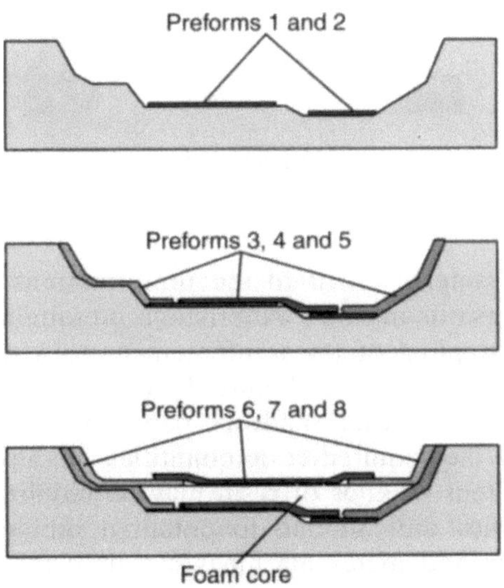

Figure 7.2 On large tools with flexible faces it can be convenient to use the mould tool to make the preforms in a sequential way as shown here.

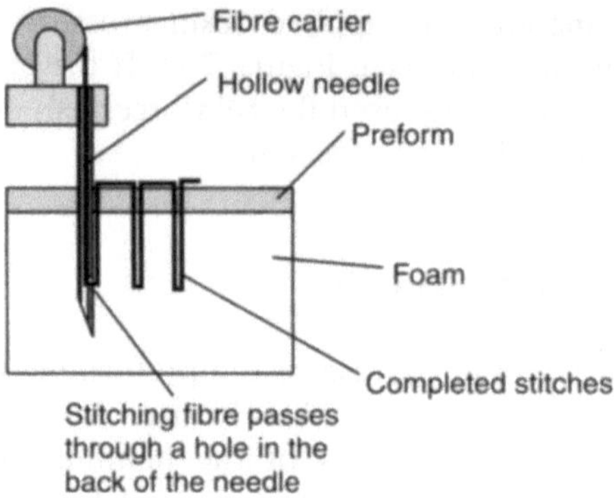

Figure 7.3 Various methods can be used to stitch together preforms or stitch preforms to foam cores, a hollow needle variant is shown here. Ideally the fibre carrier should be incremented forward as the needle is withdrawn to prevent the new stitch being pulled out. As a minimum the fibre carrier should have a low mass and friction.

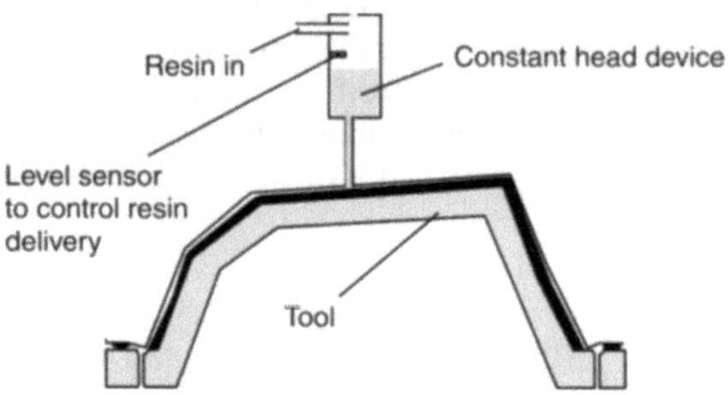

Figure 7.4 A simple constant head device can be used to ensure that resin pressure is not excessive.

water-carried systems are used the preforms must be very thoroughly dried before resin is injected. Alternative preforming techniques include stitching and stapling of the reinforcement pack. Single-sided stitching equipment such as shown in Figure 7.3 can be used to attach reinforcements to foam cores when sandwich panels are being made.

Handling of the required resin quantities can also present some difficulties. Many tens of kilos of resin may be required making the use of premix machines difficult due to potential problems with exothermic heating. If mix and meter machines are used their delivery has to be controlled such that the rate or pressure of resin delivery is not excessive. A constant head device could easily be fitted to a mix and meter machine to overcome these difficulties (see Figure 7.4).

7.3 TOOLING DESIGN FOR LARGE AREA RTM

In most cases one half of the tooling will be essentially rigid. The other half can either be semi-rigid or completely flexible. With foam cored structures it is also possible to use the foam as the only 'rigid' element and use flexible materials elsewhere. Semi-rigid toolfaces are often thin GRP mouldings. These will give a good surface finish on both component surfaces as for rigid tool RTM, but thickness tolerances will not be so well controlled. The Lotus VARI process uses this approach where the mould is clamped shut and consolidated using variable vacuum levels within the tool and between a pair of seals (see Figure 7.5). The problems of reinforcement cutting accuracy noted earlier for rigid tool RTM (section 4.3.5) are also relevant in the case of semi-rigid tooling, especially as available clamping forces can be low.

In the general case one tool element can be considered as non-rigid. This can be a tailored rubber bag made from sheet stock, or by paint or spray on techniques, from silicone rubber or polyurethane. Some light reinforcement might also be included in such bags to improve durability.[1]

Untailored bags could also be used for simpler geometries, but the necessity to stretch the bag into shape will reduce available compaction pressure, may allow bridging and may reduce bag life. Lastly, standard vacuum bag materials can be used to give a 'one shot' toolface. The resin in and out ports can be integrated into the bag or rigid tool elements, or standard vacuum bag fitments can be used for short runs. Figure 7.6 shows the general layout of such a tool.

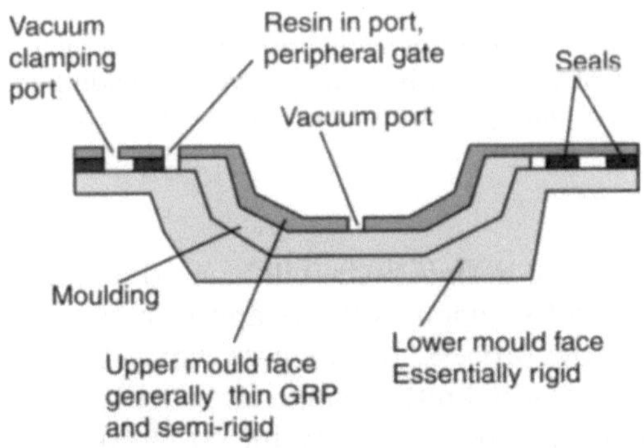

Figure 7.5 Vacuum pressure can be used to provide a clamping force on tools. Resin delivery can be by vacuum or a low pressure. If pressure injection is used with a somewhat flexible upper tool face this will deflect to speed up the injection process. Centre point injection can be used as well as peripheral injection. It has advantages if sealing is imperfect.

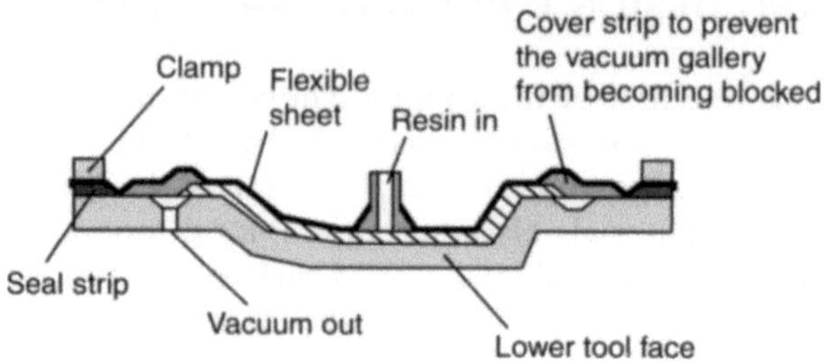

Figure 7.6 The features of flexible faced tools are shown here. The flexible sheet may be rubber or a polymer film.

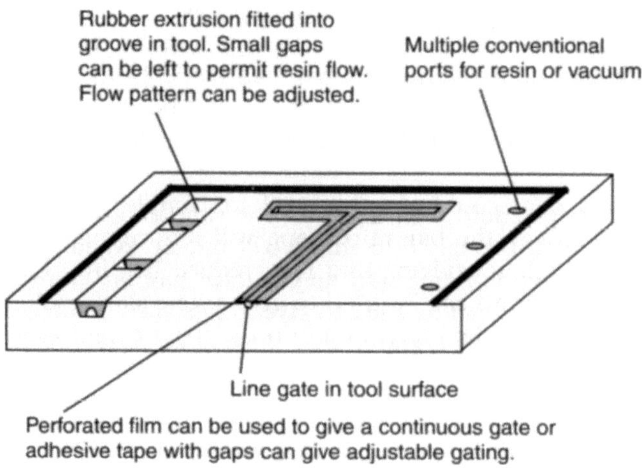

Figure 7.7 A variety of multiple gating options can be incorporated into the tool-face, some types of which can be adjusted from moulding to moulding. A similar range of multiple gating options can be incorporated into the flexible tool elements.

Whatever the tool design, resin will have to fill the tool before gelation occurs. If the reinforcement Vf% is appreciable, vacuum only injection can be a very slow process. In one case which I studied, using polyester resin and +/–45 insert knitted glass cloth with a line gate injection at 0.9 bar (resulting in >60% Vf) the injection flow length was much less than 1 m after an hour's injection. Several methods may be used to increase the effective injection rate. The simplest is injection from the outermost edge of the part to the centre, as noted in section 1.2.2. This technique is only likely to be effective for parts of moderate dimensions, dependent on reinforcement permeability. Multiple gating can be used,

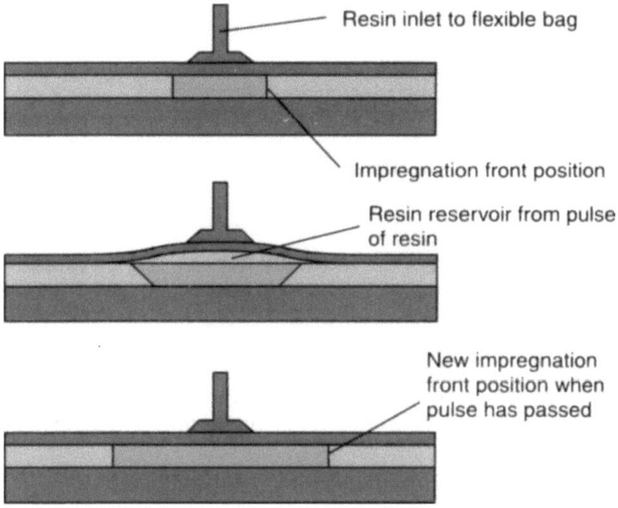

Figure 7.8 Injection rate in flexible tools may be increased by injecting a pulse of resin that bulges the bag outwards, flooding resin over the surface of the preform, as shown here.

such that the component is broken down into 'cells', with an adequate flow rate within each cell. Multiple gates could be integrated into either the rigid tool elements or the flexible elements. Figure 7.7 show some options.

In principle this approach is not size limited, although very complex gating arrangements will obviously add costs and may increase process risk. Thirdly, pulsed injection is possible. In this case resin is delivered in pulses at a rate beyond that which the reinforcement can absorb. This creates a reservoir of resin between the vacuum bag and the toolface (see Figure 7.8).

The outer edge of the reservoir passes beyond the end of the previously wet-out reinforcement such that the dry reinforcement is impregnated from the surface, rather than through the previously impregnated reinforcement. This approach can also be used with some semi-rigid tools. In a variant process the reservoir of resin can be manually distributed with rollers, prior to the delivery of the next resin pulse. I have used this approach to prepare high-quality panels very much more rapidly than would be possible without the manual assistance (minutes to fully wet out 1 m of high Vf% reinforcement rather than hours). Pulsed resin delivery could be utilized, in addition, with the other techniques previously mentioned. Another, proprietary, variant of the RTM process, called the SCRIMP process utilizes a layup as shown in Figure 7.9.

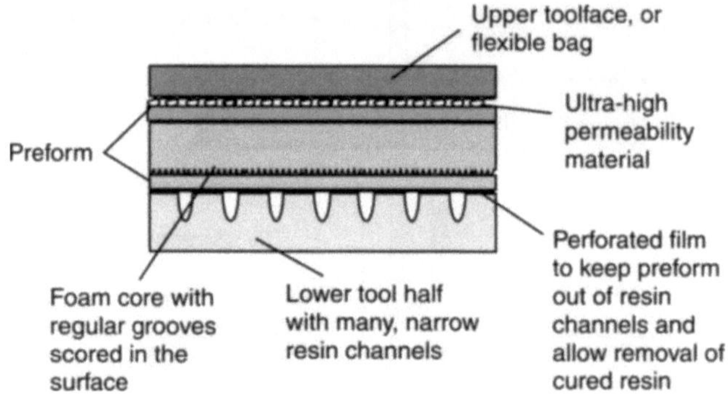

Figure 7.9 Any or all of the techniques shown here can be used to speed resin flow in large mouldings with high Vf% preforms or vacuum-only injection.

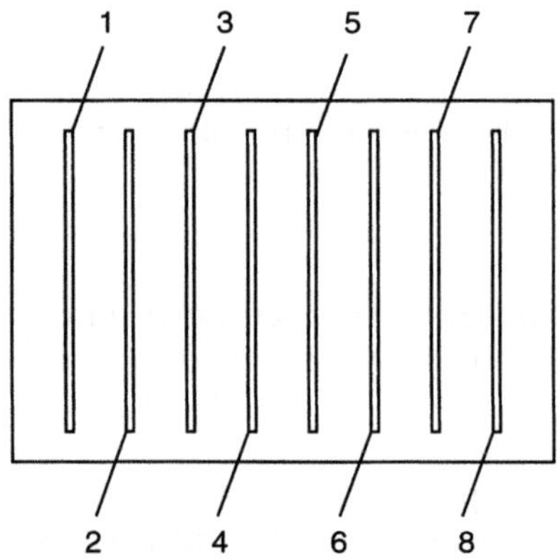

Figure 7.10 Sequential gating can be used to minimize many of the potential problems with very large mouldings; at the cost of increased complexity. Initially gate 1 is used as the injection gate and gate 2 as the vacuum gate. As the resin reaches gate 2, gate 1 is closed and gate 2 becomes the resin in gate. Gates are then switched sequentially until resin is emerging from gate 8 and the mould is full. The resin injected at gate 1 may have cured at this point.

A perforated film is used above the surface of the reinforcement layup and above this is a layer of ultra-high permeability material. The resin rushes through this layer and impregnates the reinforcement by lateral flow. In all the cases noted above the resin delivery machinery must be sized to the maximum resin flow rate demanded, either continuously or as a pulse. Lastly, if multiple, independently controlled, gates are provided, sequential injection is a possibility. In this case each cell is effectively an independent tool and can be filled by opening the relevant gates (see Figure 7.10).

Despite the obvious increases in complexity there are two potential advantages. Firstly, the resin injection machinery only needs to be sized to the demands of one cell so that premix machinery could be used. Secondly, the cell size can be kept small enough that more rapidly curing resins may be usable if the component design requires the use of specific resins with rapid cure characteristics. In principle, the use of sequential injection relieves all of the size limitations due to injection machinery or cure-time characteristics.

Designing, manufacturing and handling very large tools will tend to be more difficult than for small tools. The reasons for this include weight, possible complex resin porting requirements, special ejection requirements for large parts, the difficulties of acquiring master models etc. In most cases the flexibility of composite tool manufacture will lead to this tooling route being followed. To obtain adequate rigidity in the tooling the use of foam-cored sandwich panels may be required. Some high-speed vessels are now being made by constructing a series of decks and bulkheads and assembling a foam core onto these. The details of the foam

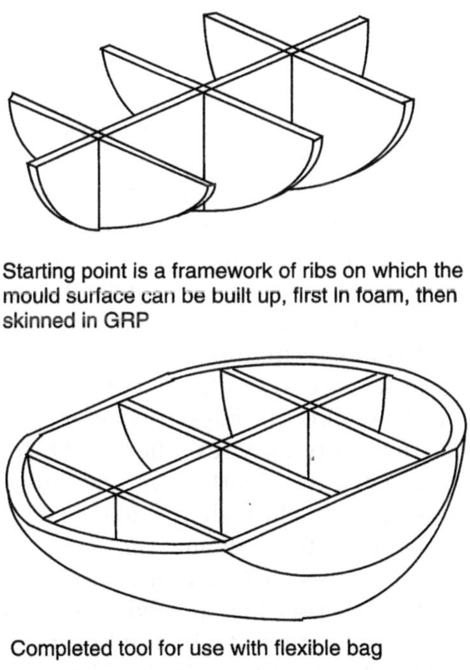

Figure 7.11 Very large tools can be made by techniques similar to those used to make sandwich-cored ships.

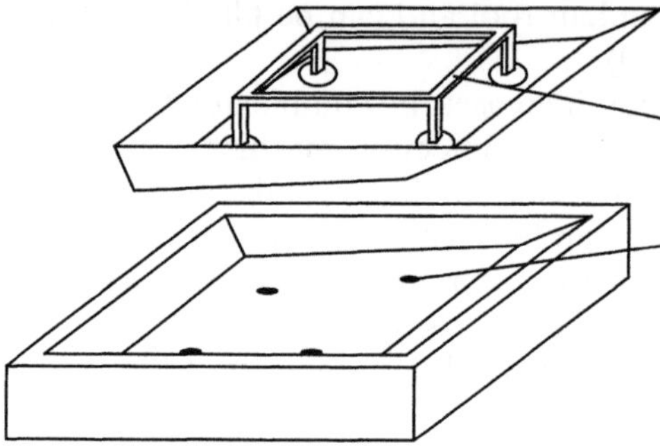

Figure 7.12 Vacuum cup manipulators can be used to eject and transport large mouldings.

core are then machined prior to laminating on the outer surface of the composite skin. The internal structures are then removed and the inner skin is laminated in (see Figure 7.11).

A similar approach could be used to make very large RTM tools without the necessity for a master model. Ejectors could utilize air ejection for initial separation with vacuum cups to grip and lift the component free (see Figure 7.12).

However tools are made, the costs to acquire an adequate very large RTM tool can be expected to be high. Very large parts are seldom required in high volumes and it may be impossible to justify the production of RTM tooling. One way out of this dilemma for foam-cored sandwich panels may be to utilize the foam core itself to act as the RTM tool. In this case the outer 'tooling' would be expected to be a plastic film, which could be polyethylene, a stretchable film or standard nylon vac-bag materials etc. For large-area mouldings it may be necessary to joint the film, and standard mastic tapes can be used for this; they can also be used to seal the bag to the foam. One problem with joints in thin bags (and also with wrinkles in bags) is that they can act as easy resin flow paths and disrupt an orderly resin flow front. Resin in and out gates can be of the forms discussed earlier. In addition it is possible to port resin into the tool directly through the foam (see Figure 7.13) and so avoid having to penetrate the vacuum bag.

In the absence of a resin distribution network the rate of resin flow would be expected to be very low at moderate to high Vf% levels. The sort of techniques noted earlier can be used to increase flow rate, as can the use of a striated foam surface. If the film is directly sealed to the foam the resulting edge will have to be cut away. A separate edge closure could be used to avoid this if required (see Figure 7.14).

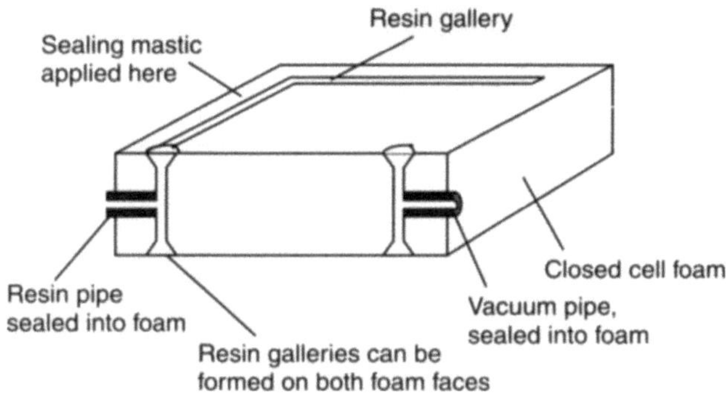

Figure 7.13 When using foam cores with large components it is possible to use the foam as the tool. Resin galleries can be cut directly into the foam and a vacuum film sealed with mastic strip. This method uses more consumables than a separate tool, and trimming of the moulding is required, but tooling costs are saved.

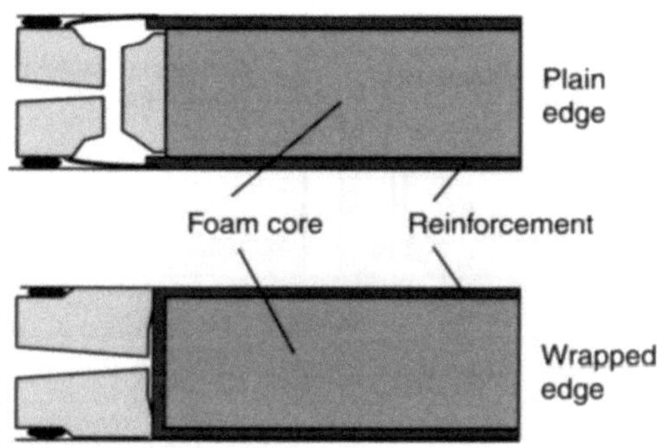

Figure 7.14 As an alternative to porting the resin through the foam a separate edge strip may be used.

In any case the use of a double vacuum seal is recommended to ensure vacuum tight bags which are critical to the process (see Figure 7.15). Lastly, it is not uncommon for small RTM tools to require several moulding trials before the details of the process are under control and reliable production is achieved. In the case of large mouldings, and especially one-offs, this is clearly unacceptable. Obviously, great care will have been taken in planning injection and gating strategy; despite this, uneven flow could easily lead to some air entrapment and the potential for scrap components. For this reason 'emergency' porting arrangements should be provided so that local dry spots can be eliminated if required.

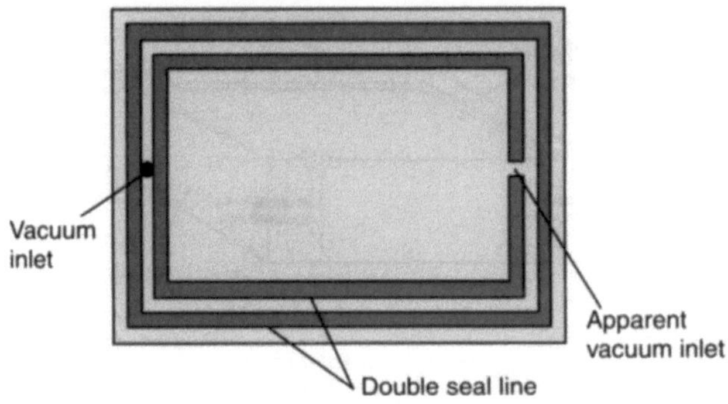

Figure 7.15 When using vacuum to consolidate and drive the resin flow perfect vacuum sealing is very important. Imperfect seals will draw air into the tool cavity, double seals can prevent this. Any air that penetrates the outer seal will be drawn straight to the vacuum inlet rather than into the tool. In addition to ensuring that no air is introduced this technique can be used to give more flexibility on the position from which the vacuum is drawn.

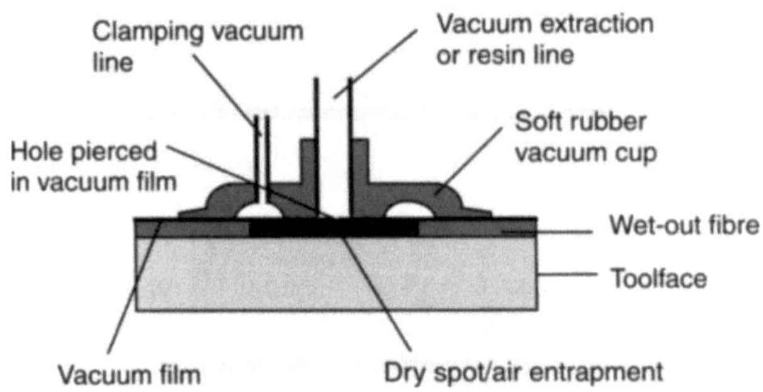

Figure 7.16 'Emergency' porting arrangements as shown here can be used to correct flow problems during injection into tools using a polymer film as one tool face.

Figure 7.16 shows a possible design for such emergency arrangements, mastic seals could be used in place of the vacuum cups shown. The development of reliable emergency porting arrangements would also permit complex porting arrangements to be more easily achieved.

Component design must also be modified to take account of the differing characteristics of very large components. For example, complexity is likely to be lower, draft angles higher and undercuts avoided. For very large components made by RTM a substantial investment in materials and labour will have been made before injection commences.

The control of process risk in the injection is therefore of paramount importance and all component and process design efforts should be focused towards the reduction of this risk.

REFERENCE

1. Musch, G. and Bishop, W. (1992) Tooling with reinforced elastomeric materials, *Composites Manufacturing* **3**(2), 101–11.

8 Thick section RTM

This area has probably not received as much attention as some other RTM variants, but RTM is being developed or considered for components such as heavy armour, heavy leaf springs and other highly loaded components. The limits of current capability are not well defined, but I have been involved in the moulding of components 2 m long with a maximum thickness of 50 mm and solid blocks 80 mm thick. Heavier section parts than these may well have been developed by other investigators.

The difficulties in making thick RTM components fall into two categories; those that are relevant to any and all thick composite components and those that are process-specific. In the first group are high-cure exotherms, ensuring cure uniformity, thermal and cure shrinkage stresses (especially in curved or complex parts) and ensuring that allowable stress values determined on small test pieces are relevant to the large volumes of material that may be present in thick components. The handling of these potential problem areas is generic to any thick component composites work and will not be discussed in detail here. Thermal modelling will almost certainly be required, both for peak exotherms and induced stresses, as these latter can be a substantial proportion of available out-of-plane properties. Resins that are slow curing at relatively low temperatures are likely to be preferred, as are those that exhibit little cure shrinkage. Low-temperature curing epoxies would seem to have the right sort of properties. The question of relevance of allowable properties has two elements. The first is to be certain that the defect spectrum in the small test pieces is the same as that found in thick components. That is to say that if some defect types are only seen in thicker sections the use of thin test pieces cannot realistically predict the properties of the large volumes of material. The second is that, even when the defect spectrum is the same, the probability of finding more strength reducing defects is higher for the larger volume (both in the senses of finding more defects and of finding individual defects that reduce the strength by a larger factor). In principle the use of a Weibull statistics approach can be used to predict the scale of the effects,[1] but there is considerable uncertainty about the extent to which these approaches yield fully valid data. In practice it is probably better to try to develop design data at a scale which is not too far away from that of the parts to be made.

The RTM specific problems revolve around the control of the resin flow front as injection progresses. In most RTM work the permeability can be assumed to be a two-dimensional function, as the flow distances in the through thickness direction are small. For thick section work this is not necessarily the case and 3D flow modelling may be needed. It seems likely that it is not the absolute thickness that is critical, but the ratio of thickness and other dimensions. For example, controlling the flow front should be easier in a large, thick, square plate than in a long and narrow part of the same thickness. In the second case the ratio of mould edge length to cavity volume, and hence the likelihood of easy flow paths is much higher. The worst case would be that of a long, narrow, thick, UD reinforced part of high Vf%. In this case the through thickness permeability can be very low, whereas longitudinal flow can be rapid and edge flow would be expected to be very rapid. In these circumstances it is very easy to entrap air as the injection proceeds. Flow inhomogeneities caused by mismatches between tool and component geometries around radii can also be very serious for thicker parts. For planar components, remote from injection points, stable flow fronts should be relatively easy to arrange. Even in this case, sudden changes of section thickness can be expected to be potential problem areas.

Many of the approaches that can be taken to ease the problems are similar to those discussed earlier. Easy paths should be minimized, or they might become part of the injection galleries if outside-in injection is used. Very accurate preforms can help to minimize easy paths at mould edges and section changes, but a consideration of likely production tolerances on thick preforms makes it likely that the control of easy paths will be more difficult as thicknesses increase. One approach to problems of easy flow paths at mould edges is to introduce a very compliant foam edging strip to take out much of the difference in size between preform and tool, mastic materials have also been used to fill easy flow paths, but such techniques would need considerable development to be reliable in production.

If it is not possible to eliminate easy paths, then the emphasis must be on maximizing the preform permeability to minimize the influence of the easy paths. If low-volume fractions are structurally adequate this is the simplest option. Lower Vf% can also be simulated by such devices as injecting resin into moulds that are sealed with inflatable 'O' rings and held slightly apart during injection. At higher volume fractions, techniques such as those discussed in section 2.1. can be of benefit to reduce the

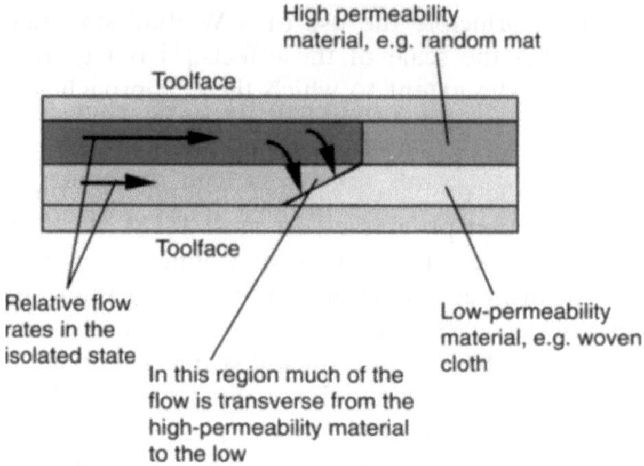

Figure 8.1 When high and low permeability materials are used together most of the bulk flow goes via the high permeability material. Even with very different permeabilities a stable flow front is quickly established.

resistance to flow. In addition, techniques such as using sandwiches of high and low Vf% material can be used (see Figure 8.1).

The critical factor in some cases (e.g. where resin has to be ported to a surface gate in a thick part) may be through-thickness permeability. Some of the reinforcement designs for increased in plane permeability may also be effective in increasing transverse permeability, but it is by no means axiomatic that this would be the case and specific measurements of transverse permeability may be required. The use of holes formed through the preform at the in gate (formed with a spike prior to injection) has also been used to ensure that the initial wet-out through the component occurs without air entrapment. In addition the use of evacuated tooling will greatly ease the production of good components. However, vacuum levels may have to be very high indeed to ensure adequate quality in the absence of adequate flow control. As for large area RTM the requirements for the transfer of large amounts of resin will have important consequences for the design of resin injection machinery.

Lastly, very thick mouldings could be sequentially produced. A moulding of dimensions that can reliably be processed can be made and prepared for bonding. A further preform can be applied to this moulding and impregnated with resin. (The use of a film adhesive at the interface would be advantageous.) The process could be repeated until the required thickness were achieved. There are obviously cost increases with this approach that would have to be set against reductions in process risk, in some cases there may be overall benefits available.

REFERENCE

1. Wisnom, M. R. (1991) The effect of specimen size on the bending strength of unidirectional carbon fibre-epoxy, *Composite Structures* **18**, 47–63.

9 Known applications of RTM processing

The use of RTM as a processing technique for advanced composites is relatively recent. A listing of applications such as that given below is thus inevitably going to be incomplete due to the pace of developments. Despite this it is thought to be worthwhile to attempt to indicate the breadth of applications explored to date. RTM is one of the few composites processing techniques used for both high-performance aerospace applications and larger volume industrial applications. For this reason both sorts of application are noted below. The absence of any particular product type should not be taken as being indicative of the unsuitability of RTM in that case; merely that to my limited knowledge it has not yet been explored.

9.1 AEROSPACE AND DEFENCE

1. access covers and doors
2. control surfaces
3. de-icing duct components
4. drive shafts
5. electronics cassettes
6. engine cowling beams
7. fan blades
8. fins and wings
9. fuel tanks
10. helicopter drive shafts
11. I beams
12. infra-red seeker housings
13. launch tubes

14. military equipment boxes
15. missile bodies
16. precursors for carbon/carbon component manufacture
17. propellers
18. radomes
19. rotor blades
20. shaped armour panels
21. stator vanes
22. space station struts
23. sub-munition dispensers
24. thrust reverser components
25. torpedo hull and underwater weapon prototypes

9.2. AUTOMOTIVE USES

1. body panels and body shells
2. bumpers
3. cargo van roofs
4. clutch and gearbox housings
5. chassis cross members and components
6. front and rear chassis sections
7. leaf springs
8. pick up truck boxes
9. tub chassis with integral floor pan
10. space frames

9.3 CONSTRUCTION

1. columns and posts
2. commercial building doors and frames
3. in-situ beam strengthening
4. kiosks and doors
5. manhole covers
6. restaurant fronts
7. signs

9.4 ELECTRICAL AND ELECTRONIC

1. business machine housings
2. communications and transmitter housings
3. computer work stations
4. parabolic dishes
5. radomes

9.5 INDUSTRIAL AND MECHANICAL

1. cooling fan blades
2. compressor covers
3. corrosion resistant equipment
4. drive shafts
5. electrostatic precipitators
6. floor boards
7. flywheels and flywheel system components
8. gearboxes
9. inspection hatches
10. mixing blades
11. protective helmets
12. RTM machine housings
13. solar reflectors
14. tooling bars
15. gate valves

9.6 MARINE

1. boat hulls
2. cabin covers and components
3. decks
4. dockside power posts
5. emergency escape equipment and housings
6. propellers
7. radar bridges
8. submarine masts

9.7 SPORTS EQUIPMENT

1. amusement rides
2. bicycle frames and handlebars
3. golf carts
4. golf club shafts
5. jet-ski bodies
6. kayaks
7. public area furniture
8. sailboards
9. skateboards
10. surfboards
11. swimming pools

9.8 TRANSPORTATION

1. barge covers
2. hopper car covers
3. light rail car and monorail doors, car and chassis components
4. trailers, including insulated variants
5. truck cab components, e.g. bumpers, hoods, doors, floor pans and sleeper cabs
6. truck wind deflectors and aerofoils

Over 80 classes of components are noted above, in many very different markets. No better testament to the flexibility inherent in the many variants of the RTM process could be asked for.

10 Troubleshooting RTM processing problems

Problems can occur in RTM processing during all stages from tool tryout, through prototyping and on into production. In some ways it is easiest to work with production problems as in this case there is at least the assurance that the process has at some time been working as required. This case will, therefore, be considered first. The initial step is to quantify the problem, identify the required behaviour and the departure from the required behaviour. The critical point here is to ensure that the problem is real. For example a change in inspection staff may have led to a change in reject rates, even though the absolute quality levels are unchanged. Problems are usually defined through a change in product quality, but other problem types are possible such as an increase in downtime of injection machinery or damage to tooling.

For example, if constant flow rate mix and meter machinery is used without pressure limiters and the resin viscosity increases for some reason (or the reinforcement permeability is reduced) the machine may well continue to inject resin at the same rate, possibly leading to no appreciable change in process time or product quality. On the other hand the driving pressure will have risen substantially with effects on seal life or linkage wear in the dispensing machinery. Also there may be effects on resin mix ratio (if seals start to leak) or on fibre washing, or in the worst case tool damage may occur. In this case examining process records would give but little indication of the problem's root cause. If viscosity checks of incoming material were made and recorded the problem would be simple to diagnose. In the absence of these records diagnosis would be much more problematical as the presenting symptoms require more interpretation to reach the root cause. The first step to an easy resolution of problems can thus be seen to be in the keeping of adequate and relevant records of both the incoming materials properties, the process

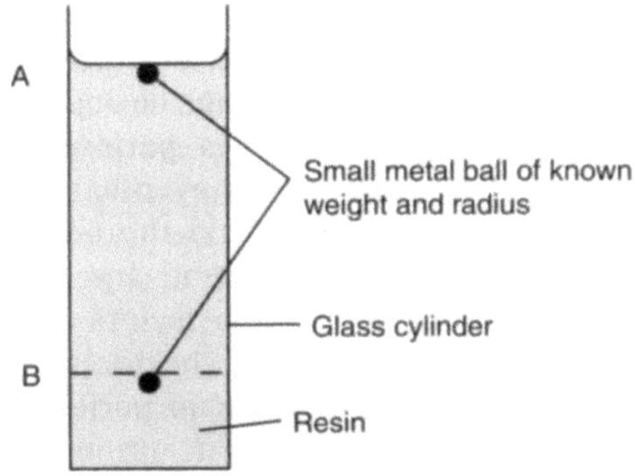

Figure 10.1 Simple measures of viscosity such as cup viscometers or falling ball types (shown here) are useful to control the quality of incoming materials.

parameters, and the required product quality. This is often seen as excessively burdensome in smaller factories, but very simple checks can often yield useful results. For example falling ball viscometers may lack the sophistication of more common types, but cost very little and require no great skill or time in use (see Figure 10.1).

Cup-type viscometers, where the time taken for a cup of resin to empty through a small orifice at the bottom are also very simple in use.

Assuming that a check of available records cannot detect any sudden variations that could explain the changes in process or product quality, the second stage is to monitor the process in operation closely, using the manufacturing instructions and process control paperwork as a guide to the procedures that give rise to stable performance. As before, unless such records are available and adequately detailed the resolution of problems will be more difficult. This monitoring needs to encompass all stages in the production from reinforcement purchase, storage and preparation, to mould loading, injection, cure, demoulding and deflashing. As noted above it is not axiomatic that a problem that appears to be created during resin injection has its roots there. It is, of course, absolutely necessary that the manufacturing instructions do indeed reflect those activities that go into the generation of good parts and a stable process. This may not always be the case if manufacturing plans have been drafted remotely from the part prototyping exercise. If this proves to be the case, and those involved in the prototype developments are unavailable, then one can only proceed as one would for a new component development.

Lastly, if all aspects of the process seem in be in line with relevant and carefully produced manufacturing instructions a more thorough study would be indicated. This needs to include checks of the process equipment, especially quality sensitive items. In one case known to me, product quality had slowly declined to an unacceptable level. A study of the line showed that materials were correct and that procedures seemed to be being followed. A process study, varying the injection parameters over wide ranges, led to an improvement in performance, but not to the required quality level. The problem was eventually traced to a combination of tools that could not reliably hold the necessary vacuum levels, contaminated vacuum checking equipment, and poor vacuum checking procedures. Between them the last two factors obscured the problems caused by the first factor. When new, the tools could easily hold the required vacuum levels (they either sealed perfectly or not at all), and the uncontaminated vacuum equipment would have detected major problems even with poor procedures. The problem was easily rectified by tool maintenance, replacement vacuum checking equipment and new procedures that gave a more thorough vacuum check without the possibility of false positives or contamination.

In the case of new components, troubleshooting is more difficult as it cannot be guaranteed that the problems are soluble and the net will have to be cast wider to determine both the root causes and their solution. The aim is to develop a process window that is as wide as possible to minimize the effects of the minor, day-to-day, variabilities that are all but impossible to eradicate. Probably the most common problem encountered is air entrapment or voidage within the component, followed by (in no particular order) inaccurate dimensions, poor surface finish, premature gelation, cure inhomogeneity or undercure, poor wet-out, movement of cores within the tool, resin rich zones which may crack, difficulties in extraction and delamination. Before considering the first area a brief word will be said about common causes of some of the other problems.

The most obvious cause of inaccurate dimensions is inaccurate tooling. In addition, tooling that is inadequately stiff for the combination of reinforcement compaction pressure and injection pressure can give rise to the same effect. Small pieces of moulding putty can be cured between the faces of the fully closed tool to check the mould cavity dimensions in the zero-pressure state and thus the baseline tooling geometry. This tooling geometry can then be compared with the dimensions of components extracted from the tool. It may be possible to use minimal injection pressures to bring dimensions within tolerances if the tooling is correctly dimensioned but too compliant, but the preferred solution would be to stiffen the toolface.

Poor surface finish can be caused by poor tool surface finish, or build-up or poor application of release agents. In addition, especially for polyester resins of high shrinkage, a blotchy surface finish can result if curing is not controlled. As the resin shrinks the laminate thickness reduces slightly, at some point tending to release from one surface of the tool, when this happens the surface that has released may roughen as its surface is no longer controlled by the tooling. If some areas of the moulding first release on one side and other areas first release on the other side a blotchy surface may be seen. This is not a structural problem, but surface finish may be critical. Poor surface finish from this cause can often be distinguished from that caused in other ways by a comparison of both sides of the component. If the surface finish of one side is a 'negative' of the other side (i.e. a good area on one side equates to a poor one on the other) then this is the most likely cause. Holding one face of the tool slightly hotter than the other usually solves this problem by forcing cure first on one surface.

Premature gelation can be caused by: poor metering in mix and meter machines or poor resin mixing or excessive standing time in pressure pot injectors; exothermic heating in thick laminates or where tool temperature is poorly controlled and could, in principle, be caused by chemical interactions with binders.

Cure inhomogeneities or undercure are likely to be caused by a similar set of factors. In addition some curing agents are moisture sensitive and if kept in poor conditions may be unable to give full cure. In some cases rings can be seen around resin in-gates, in this case resin mix and meter machines would be suspect with respect to seal performance and wear in linkages.

Poor wet-out would be expected to be caused by a basic incompatibility between resin and reinforcement. In the case of some bound random mat materials the resin must dissolve the binder prior to wetting out the tows, if cure is too rapid the onset of gelation may occur before this process is complete. In addition any very high Vf% spots due to poor layup or reinforcement pack/tool design may show incomplete penetration and wet-out (see Figure 10.2).

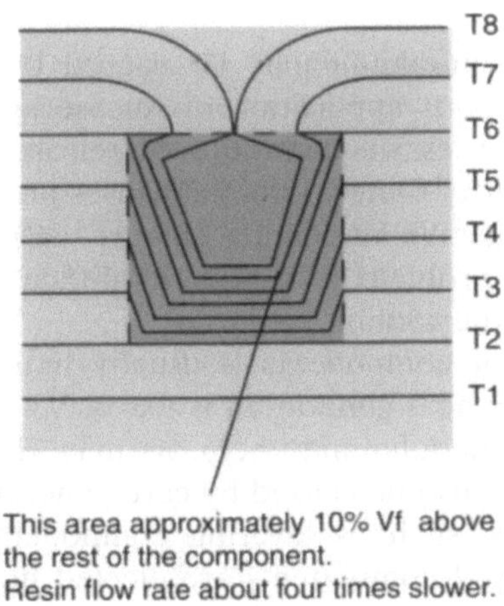

Figure 10.2 Spots of high Vf% can easily lead to areas of very poor wet-out.

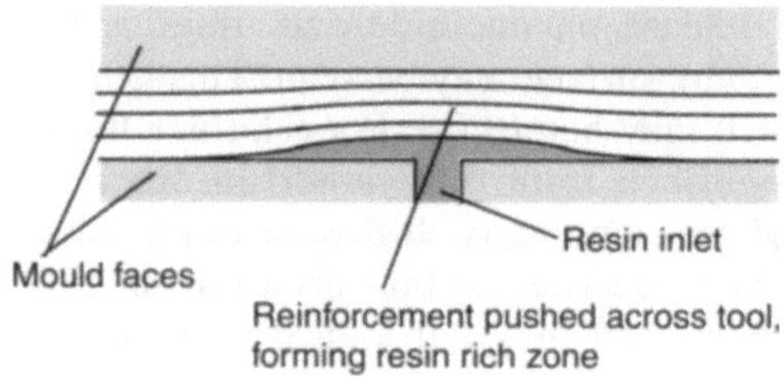

Figure 10.3 Resin rich zones at injection points can be caused by the reinforcement being pushed across the tool by the resin flow.

Movement of cores during injection is caused by flow instabilities as noted elsewhere and may be solved by venting the core to cure the instability or by ensuring that the clamping pressure across the core is greater than the local resin pressure. The gross movement of reinforcement as 'fibre wash' is also caused by inadequate clamping pressure with respect to injection pressure. Classical fibre wash only normally occurs with random mat reinforcements, similar effects can be seen with woven cloths in the thickness direction where the resin is gated in at the surface of the cloth.

Resin rich zones may be caused by a mismatch between tool and reinforcement geometries, as noted earlier (see Figure 2.6). They can also be caused by poor mould loading (e.g. preforms not 'bottoming out' in ribs, Figure 6.4) and by fibre wash or through-thickness movement of reinforcements at gates (see Figure 10.3).

Changes to preforming approaches, mould-loading procedures, resin injection pressure profile, Vf% or reinforcement type can be used to cure these problems.

Problems in demoulding may be caused by the wrong selection of release agent or poor application of release agents. It should be noted that the use of excessive quantities of release agent can lead to poor release performance. Other potential causes are inadequate tool stiffness that leads to effective undercuts (Figure 4.2) and inadequate ejector design. This last problem may be very difficult to solve without major modifications to the tooling.

Delamination of components is usually limited to thick components where exothermic heat generation is excessive. If this is the case the use of slower cure schedules may help, as may changes to the resin used. Delamination can also be caused by cure generated volatiles and by poor ejection practices, such as levering components out of tools on thin flanges, especially if components are ejected hot or in a state of partial cure to minimize cycle times. It is often assumed that partially cured matrix resins will be relatively weak but tough, as the fully cross-linked structure has not yet been formed. This is far from being true and partially cured resins can be both very weak and lacking in toughness.

Returning to the most commonly experienced defect type, air entrapment or voidage. Several different root causes may contribute to the appearance of voids or gross air entrapment. These include:

1. flow front inhomogeneities, where the flow front completely avoids one area;
2. inadequate run-off or purging of resin after injection is nominally complete;
3. lack of compatibility between fibre and resin;
4. air entrainment in mix and meter machines or inadequate degassing in pressure pot machines;
5. air leaks in evacuated tools;
6. volatile species in the resin or generated during cure;
7. absorbed water in some binders may boil off during cure;
8. resin cure shrinkage.

1. Gross flow inhomogeneities are essentially due to either the wrong choice of gating positions or to the presence of easy flow paths (see Figure 10.4).

In either case deliberately short-shotting the tool can be of some assistance, i.e. injecting a resin which will gel before injection is complete or adjusting process parameters to achieve the same effect. Note that delivering a small amount of resin, then stopping and waiting for the resin to cure, may not be so effective, as the resin that has been injected will tend to be wicked away from the flow front between the end of injection and the onset of cure. The distinction between the wrong gating positions

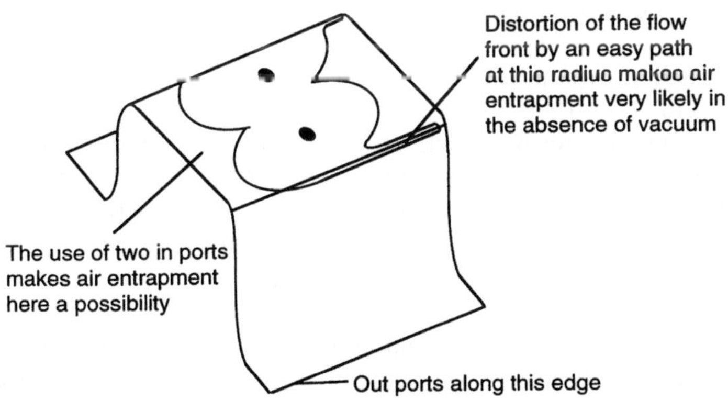

Figure 10.4 Combinations of poorly sited in-gates and easy flow paths can lead to air entrapment.

and easy flow paths is somewhat blurred as gating should be arranged to avoid easy paths if possible, but the use of short-shot techniques can give useful information. Flow modelling techniques can be of assistance in finding better injection strategies, but the ability of many modelling approaches to handle easy paths is somewhat limited. If the gating is seriously in error the only solution may be modifications to the tool, which can be very costly. The adoption of high vacuum levels in the tool may also be of benefit, but retrofitting tools for high vacuum if they were not designed with this in mind may not be possible. Reliable removal of easy flow paths can be difficult to achieve and a solution based on gating is preferable to one that relies on the elimination of all easy paths, e.g. by very accurate preform preparation and mould loading.

2. The mechanisms by which voids can form within tows of reinforcement were discussed in Chapter 1. These voids do tend to emerge from tows and travel along with the resin flow front and may be removed from the tool by resin purging. If injection ceases immediately resin shows at all the out-gates it is possible that the areas of the component adjacent to the out-gates may have a high void content. If a high level of voidage is seen close to out-gates, inadequate resin purging may be the cause. This is simply investigated by increasing resin purge times. The use of evacuated tooling will also be of great benefit. Inadequate resin purging is not the only cause of voidage close to out-gates, see (8), below. A less common problem can occur if flow rates are very slow and wetting flow dominates the speed of the flow front. In this case it is possible to form voids between the tows. This form of air entrapment would be more difficult to distinguish from that caused by resin shrinkage and would tend to be seen in similar areas. Increasing the injection rate into the pressure controlled rather than wetting controlled regime should eliminate this effect, and can be used as a diagnostic tool.

3. Lack of compatibility between fibres and resins is not a common problem. It can sometimes be seen between glass fibres and acid-catalysed phenolic resins. Lack of compatibility shows up as incompletely wet-out tows and is easily reçognized and distinguished from other types of air entrapment or voidage. Changing injection parameters may help (i.e. reduce flow rate to allow more time for wetting), but the only real solution is to change the resin or the fibre's surface treatment.

4. If air is present in the material being injected there is every chance that the resulting components will be voidy. The worst case is not the injection of large bubbles of air, as these can often be purged away, but the injection of a regular stream of small air bubbles. Entrained air can be distinguished from voids generated by the mechanisms in (2), by the

fact that the air is predominantly found between fibre tows and tends to be more uniform across the component, rather than being concentrated by the out gates. Also, increasing purge times will not affect the quality if the problem is air entrainment. Problems with entrained air may also be made worse by high exotherm temperatures expanding the air bubbles. If the voidage level changes with speed of cure (and hence peak exotherm temperature) the presence of entrained air may be suspected.

5. Leaks in evacuated tools can lead to air being drawn into the tool, especially in low pressure or all-vacuum systems such as might be used with flexible tool RTM variants. This problem is most likely to be associated with peripheral injection or the use of edge gating. As in (4), the air is predominantly found between the fibre tows and is not concentrated at the out-gates, nor is it affected by purge time or volume. In translucent GRP components this form of air entrapment is often clearly visible as regular lines of voids, which helps to diagnose the problem. The only cure is effective sealing. The use of leak-back testing, i.e. sealing the evacuated cavity and watching for a decay in vacuum level is also a recommended practice.

6. Volatile species in the resin are seldom a major cause of voidage, except in the case of cold-cure phenolics and in this case they must be regarded as a natural feature of the resin. A very uniform distribution of extremely small voids that showed no tendency towards preferential siting with respect to resin gates or with respect to intra-tow and inter-tow positions would tend to indicate this mode of voidage generation. Very high exotherm temperatures might also lead to some boiling off of volatile species. As in the case (4) above the voidage level would be expected to increase with rate of cure. In this case the excess voidage would be more likely to rapidly increase above some triggering peak exotherm temperature, rather than increasing monotonically with temperature.

7. Some species used as binders, e.g. PVA, can absorb large quantities of water. If these are injected below 100 °C and cured at higher temperatures the water can boil and generate local voidage and poor surface finish (e.g. if PVA bound felts are used as surfacing tissue). Local voidage found in regions where PVA or similar binder materials are used might indicate that this was a problem. The solution is simply to ensure that the preform is thoroughly dried prior to use.

8. When resins cure they generally exhibit some shrinkage. It should be noted that even epoxy resins can show considerable cure shrinkage, this shrinkage can be obscured with hot cure systems as the expansion of the liquid resin is very much higher than that of the solid resin, leading

to an apparently low shrinkage (see Figure 1.17).[1] If the mould cavity is unpressurized during cure the liquid resin shrinkage can lead to cavitation and the generation of voidage. This tends to be seen in resin rich zones, and in the absence of these between tows rather than within them, especially at tool surfaces. Resin shrinkage generated voids will be primarily distributed close to open ports, both inlet and outlet. This feature can help to distinguish them from voids caused in other ways that tend to be seen more at outlets. Resin shrinkage generated voids tend to be most common in very rigid tooling as the decay of resin pressure when injection pressure is removed is more rapid than in less rigid tools (see Figure 4.1). These problems can often be solved by curing the tool with a slight residual pressure, which can sometimes be achieved by simply sealing the in-and-out gates during cure if injection has taken place below the cure temperature.

Lastly, evacuating tools is often seen as a panacea for the removal of voidage and it certainly helps greatly in the removal of some problems.

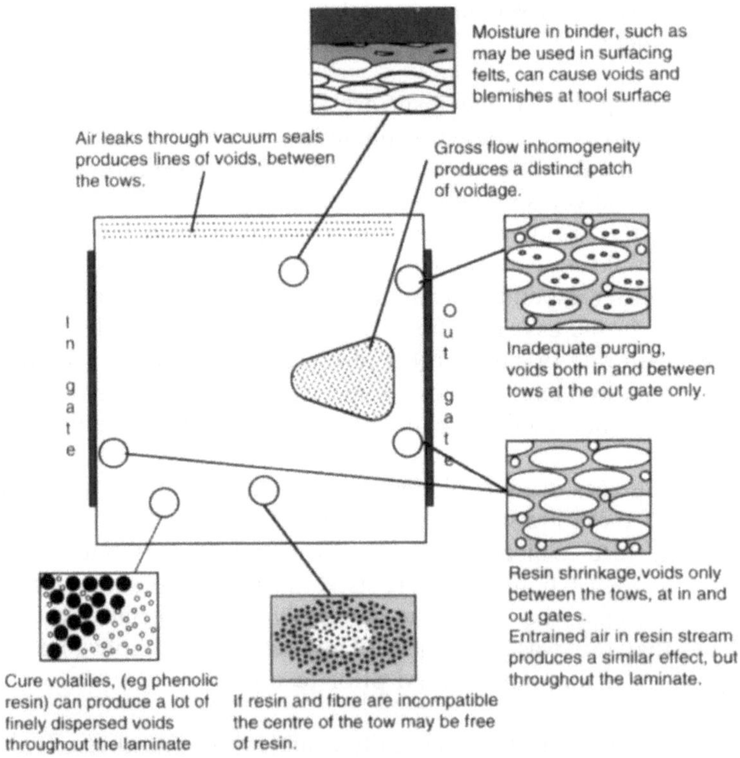

Figure 10.5 Various mechanisms may lead to voidage, the mechanism can often be identified by the shape and distribution of the voids.

However, air or volatiles that are drawn into or created within the tool, e.g. via causes (4), (5), (6) and (7), will be less affected by the vacuum and vacuum assistance has no effect on cause (8). If the air bubbles are expanded by the vacuum, the use of vacuum may interact with the resin shrinkage to make the problem worse rather than better (as the air bubbles will collapse when the vacuum is removed adding to the deficit in total resin volume). In general, vacuum assistance is of great benefit and if the level of vacuum makes no difference to voidage levels one might suspect causes (4), (5), (6), (7) and (8), so long as the vacuum generating equipment and vacuum checking procedures have been thoroughly investigated.

Tracking down the causes of excessive voidage in RTM components can be a frustrating task, but is greatly helped by an understanding of the different causative mechanisms and their usual presenting symptoms. Figure 10.5 shows the most likely positions and appearance generated by the different causes.

Finally, a listing is given below of questions that may have to be addressed when investigating process problems in RTM components. For each question the reason for its being asked is also given. As process variants can be quite different in terms of their controlling features the listing below is not in order of importance nor the order in which the questions ought to be asked for any particular case.

1. *What is the absolute vacuum level in tool, measured at the tool not the vacuum pump?*

If the flow front is poorly controlled due to easy flow paths the absolute vacuum level can be a controlling process parameter.

2. *What is the vacuum leak back rate?*

A good vacuum level produced by attaching a large pump that overcomes any leaks may still lead to air entrainment, if tools are to be evacuated they should be sealed.

3. *What is the fill time and how does it compare to predictions or previous experience?*

Sudden changes in fill time may be due to materials changes or malfunction of resin delivery system.

4. *What is the pressure at the resin supply vessel or pump, including any cycling effects?*

If mould fill times vary from prediction or experience the assumed delivery pressures may be incorrect for a variety of reasons.

5. *What is the resin delivery rate with the tool disconnected?*

The resin delivery rate should be controlled by what is happening in the tool, rather than restrictions elsewhere in the system. This is only likely to be a source of problems in pressure controlled injectors, unless volume controlled injectors are seriously malfunctioning.

6. *What is the pressure drop in any resin delivery pipework?*

If the delivery pipework has been changed, or crimped or partially blocked this can lead to greatly reduced flow rates.

7. *What is the total air, moisture, volatiles content of the resin as supplied to the tool?*

Any of the above can lead to voidage.

8. *What are the resin temperatures, associated viscosity and open times at the resin pump, tool inlet, tool cavity etc.?*

Discrepancies in these can lead to slow injection, premature cure etc.

9. *For mix and meter pumps, what is the gel time at the back pressures seen at various parts of the injection cycle, does this change at different points in the stroke for linked cylinder machines with fixed meter ratios?*

If the gel times vary with back pressure or at different points in the stroke then mix ratio is varying and seals and link wear would be suspected.

10. *What is the load required to close the tool, does it correspond to the required Vf% in the reinforcement or previous experience?*

Low loads indicate insufficient reinforcement to fill the tool completely, perhaps leading to inhomogeneous flow or fibre washing. High loads indicate tight spots in the reinforcement pack, changes in reinforcement weight, or perhaps point to insufficient softening of heat-softened binders.

11. *Are there any changes in binder size distribution, binder content, or binder chemistry?*

Any of the above could cause changes to the process parameters.

When vacuum levels or resin pressure or temperature are being measured as part of a process investigation, the normal sensors used in production should not be used. The problem may lie not in the process itself but in malfunctioning sensors that are pushing the process beyond its limits while giving the appearance of acceptable process parameters.

REFERENCE

1. Yates, B., McCalla, B., Phillips, L., Kingston-Lee, D. and Rogers, K. (1979) The thermal expansion of carbon fibre reinforced plastics. Part 5. The influence of matrix curing characteristics, *J. Mat. Sci.* **14**, 1207–17.

11 Suggestions for good practice in the design and development of RTM components

The suggestions made here are mostly specifically aimed at RTM manufacture, although some may be of a more general nature. The listing is intended to be no more than an aid in the generation of sound and reliable designs that are suited to RTM manufacture. In any particular set of circumstances the suggestions made here may be in error due to the specifics of that case. For this reason they should not be seen as either prescriptive or proscriptive.

1. The successful use of RTM requires that full account is taken of the process strengths and limitations in the entire design process. Do not erect artificial barriers between design, development and production functions.
2. Do not assume that the minimum weight solution is any sort of optimum design.
3. Be careful that added complexity does not lead to cost or quality problems when many parts are integrated into a single RTM component.
4. Do not be afraid to prototype on soft tools, so long as the differences between soft and production tools are well understood.
5. Design the part and all the associated tooling, jigging and shop floor layout taking full account of the production volume. Approaches that are ideal for one-off or small volume production may be completely wrong at high volumes.

6. Never skimp on the generation of manufacturing instructions, these need to be as complete as possible. Equally in-process inspection and quality control are likely to keep costs under control by ensuring quality.
7. Thorough interspersing of different ply orientations can limit some internal stresses. On the other hand 3D preforms made from cloth packs in which there are various ply orientations are more likely to wrinkle than if packs of a single orientation are used; as a greater degree of in-plane shear is required between plies.
8. Production drawings must be representative of the parts actually being made. If they are made too early in the design process any modifications made to ensure manufacturability in development will lead to drawing changes; adding costs and leading to the possibility of the wrong drawings being used in production.
9. If a design uses intersecting ribs be sure to allow for adequate fibre continuity at rib intersections. 3D stress analysis of intersecting ribs is often required with hot-curing resin systems.
10. Ensure that cutting patterns for reinforcements make the best use of the available space. It may be better to split plies to improve the reinforcement utilization figures.
11. Ply splices should always be butt jointed and not overlap jointed in thin components.
12. If quality control levels must be high provision should be made for in-process testpieces on cutting patterns and in the mould design.
13. Be very careful of thermal stresses when moulding in metallic inserts. Post-bonded inserts may well be stronger.
14. Removable slave bushes can be moulded in to provide seatings for post-bonded items with minimal post-moulding operations.
15. Any disposable items used in production should be costed into the part, rather than treated as an overhead cost.
16. In most cases there is no point in trying to squeeze every last minute out of a production cycle, unless the labour released can be effectively used elsewhere.
17. Preforming requirements, capabilities and limitations must be carefully considered at the earliest point in the design stage, and not left until the design is fixed. Designs are essentially fixed in their major characteristics (including production costs) by the end of the outline design phase; therefore the preforming considerations must form part of the conceptual design activities.
18. Always consider draft angles and tooling lock-ons carefully. RTM can be used to generate complex, parallel sided or undercut components, but achieving these features tends to add tooling and labour costs.
19. Sometimes (e.g. in thin ribs) large draft angles can lead to resin rich zone formation at the rib root unless the preform is made very

complex. In these cases there may be an advantage in using small draft angles and root radii. Each case is different and needs to be addressed on its own merits.

20. Do not automatically choose the highest possible Vf% in RTM, even with highly stressed parts. Lower Vf% allows faster injection, more flexibility in preform design and the process window is likely to be less sensitive to local Vf% variations.
21. Design tool heaters to give a heat-up rate appropriate to the production rate envisaged.
22. Very complex preforms may be better broken down into simpler preforms to reduce the total process risk.
23. For moulds that must be vacuum sealed it is much easier to achieve a good seal on a flat plane.
24. Tools and preforms should be designed so that the preform cannot hold the mould open, and that any loose fibres at preform edges cannot bridge seals.
25. Do not use foam cores unless it has been demonstrated that the core does not fill with resin at the injection pressure.
26. Adequate resin input and output galleries must be provided. Round bottomed 'V' grooves are easier to clean than rectangular or semicircular cross-sections.
27. It can be easier to remove thick flash from a tool than very thin flash, as this breaks up when removed from the tool and tends to cling to the tool surface. Easily cleaned moulding edges and tools can be achieved by using a narrow flash gap that connects to a deeper resin groove.
28. Pressurized resin that is expected to penetrate between fibres a few microns in diameter will penetrate down any channel within a tool, no matter how narrow. If the resin should be excluded from some area within the tool this feature must be designed in, rubber blocks have been used successfully for this duty.
29. Always consider ejection requirements carefully and ensure that adequate ejectors are provided.
30. If 'O' ring seals are used the seating groove will fill with resin on one side. Square section rings can be easier to clean, or 'O' rings can be fitted into rounded off grooves rather than the conventional square section groove.
31. Design resin in- and outlets for ease of cleaning. The use of throwaway connections or pipework can, in some circumstances, be cheaper than the labour involved in cleanup operations and solvent use may be reduced.
32. The function of design and development is to generate profitable products that function reliably, anything that deflects from this target must be strenuously avoided. RTM widens the range of the available

options allowing for more cost-effective solutions in many cases. The potential for improvements will only be realized if the design and development process fully reflects the capabilities and limitations of the process.

12 Costing

The principles of cost estimation for RTM components are the same as those for any other manufacturing process. Materials, tooling and equipment costs must be calculated, direct and indirect labour requirements must be assessed and the overhead burden of the line or product must be ascertained. The mix of capital and labour requirements for an RTM line will be substantially different from that found on an autoclave or contact moulding line. This means that the common practice of simply dividing the total overheads by the number of shop floor workers to give a factory-wide burdened labour rate will not give a true picture of the overhead costs for the RTM line. Both overestimation and underestimation of the RTM costs could arise from the use of factory-wide burdened labour rates, depending on the baseline processes used elsewhere in the factory.

Three levels of costing can be defined in the development of composite products by RTM. These are also relevant to any other process, but are included here for completeness and to indicate some of the differences between RTM and other approaches. The first two are more appropriate to project specification and outline design stages and the last to detailed design and production preparations.

12.1 TOP DOWN COSTING

This approach is used at the beginning of the design process to focus down on the available options for materials and processes, as part of the project appraisal activities. For most products a target price is available, either from potential customers or generated internally by past experience. Costs do vary with time, for that reason no £ or $ signs will be seen below and the numbers just relate to unspecified units of currency.

For example, for a part with a selling price of 100/unit. A profit of 10/unit might be required, the general and administrative overheads might be 20% of sales price, or 20, leaving 70/part as materials, labour, tooling and other variable costs such as QA. If the part is expected to weigh 5 kg at a materials cost of 8/kg, and tooling 5/unit for 1000 units, then the maximum available manhour and overhead costs are 25/unit. If the plant normally operates at a partially burdened hourly rate of 25/shop-floor labour hour the maximum labour input would be one hour.

This is, of course, very simplistic and breaks some of the rules expressed earlier concerning not using standard figures for non-standard activities. On the other hand a figure of 30/unit as total labour, QA and tooling has been quickly arrived at so that the relative contributions of these within that total can be investigated.

The point here is that even a crude outline costing at the earliest stages of a project can indicate whether or not the project is viable at all, what materials might be acceptable on cost grounds and the sensitivity of the price to substitutions of capital for labour. Note that the estimates here exclude any recovery of development or other pre-production non-recurring costs.

In conclusion, the carrying out of a basic top-down costing at the earliest possible stage in a project is a most useful means of steering the design process and preventing the adoption of designs that are too costly to manufacture. I have been called in to look at more than one project at a fairly advanced stage of development where the whole financial justification for the development could be undermined by a back-of-the-envelope calculation that took no more than 10 or 15 minutes to carry out.

12.2 OUTLINE COSTING

This approach is used more to track the design process and to ensure that design details or changes do not result in additional costs that are disproportionate to the design advantage gained. Outline costing is intended to be a bottom up process in which all the elements of the cost are identified and added up to estimate the production cost, with some simplifications made to speed the process.

The simplest system is to assess the materials requirements and the labour hours to make the part, plus some scrap and rework allowance. All the overheads of the factory are then divided by the factory wide number of shop floor hours to give a fully burdened labour rate (the cost of each hour of productive labour and all the support this requires).

If data are available for just RTM line overheads it would be much better to use these data than whole factory data. As noted earlier this can give a misleading impression of the costs as indirect labour support and other overheads can vary dramatically with process type and component market. However, having decided which process and materials options to follow on the basis of outline design and top down costing, this sort of outline costing is more a way of keeping track of how design decisions change manufacturing costs during the detailing process. For these purposes fully burdened labour hours may be acceptable, although any major design changes should be assessed for changes to overheads structure as well as more obvious technical features. This perhaps applies most strongly to production engineering decisions such as levels of automation.

If a decision is required on whether to automate an RTM process the use of fully burdened hourly rates is not acceptable. The reason for this is easy to see. If the automation will save say 1 labour hour/part (estimated as 25 units of currency) for 10 000 parts/year, and the machinery costs say 500 000 to buy (amortized over four years) with a yearly maintenance and running cost of 125 000 (split evenly between consumables, power and labour etc.) a realistic costing would indicate no financial benefit even if no finance charges were applicable. If the amortization and power costs had been applied to the factory wide overheads rather than the specific line overheads a substantial saving might be apparent, whereas in fact the costs of the RTM facility were merely being picked up by other lines in the overall factory.

To estimate the cost for a new product, the obvious starting point is the materials cost. In most cases the major elements of the part weight and thus material content are close to being fixed at a relatively early stage in the design process. Changes can obviously occur at any point during the design process, but the material content is unlikely to vary greatly. Computer software is available to minimize the wastage in cutting out reinforcement plies from cloth or other reinforcements; this can also be done manually without too much difficulty. This process allows the in-process wastage to be taken into account and generates reasonably accurate figures for costs-estimation purposes.

To estimate labour hours prior to prototyping is more difficult. Historical data on the time taken to achieve certain actions may be available, but in many cases this will be incomplete or missing. In this case the total production scheme can be broken down into very small blocks, each of which can be assessed with reasonable accuracy. For example preparation of preforms by heating and cold pressing of blanks might be assessed as follows (assuming a requirement for full materials traceability, if traceability is not required then these notes can be amended accordingly).

1. Withdraw roll of cloth from stores, check and record stock level, transfer material batch numbers to component process sheets.
2. Lay out a two-ply stack of cloth of dimensions A × B.

3. Transfer to preform stack press (vacuum type) and start heat cycle. Sign off process sheet for correct heat cycle.
4. Remove stack from press.
5. Obtain templates, check against process sheet requirements.
6. Cut out stack to templates, average cutting speed *X* m/min.
7. Label preform blanks with template and material batch number.
8. Bag up, seal and return roll of cloth to stores or shop floor holding area.
9. Transfer preform blanks to preform area.
10. Insert preform blank into preform rig carrier plate.
11. Heat blank for the time specified on process sheets, *X* mins.
12. Press blank into preform and hold pressure for the time specified on process sheets, *X* secs.
13. Extract preform, inspect and transfer to kitting area with completed paperwork signed off.

Listings such as this are easily made, and may run into hundreds of actions for complex components. Even if RTM is a relatively little known process in the particular factory situation, many of the individual sub-activities will be found in any composites fabrication line. Relatively accurate figures should be available for labour consumption in these activities, reducing the area of uncertainty to novel or unusual features of the design.

Wherever possible real, measured, times should be used to avoid over-estimation due to uncertainty in the correct times to choose. Just as importantly, such activity listings can assist in identifying the most efficient working patterns. For example, in the listing given above each preform is made from the basic reinforcement by a single operator as a single unit. This may very well be inefficient compared to some sort of production line system where each operator carries out a smaller group of actions repeatedly. Without a detailed work breakdown it will be very difficult, if not impossible, to identify the most efficient working practices and the best shape for the production line. Activity listings can also be used to identify control points and inspection points for both operator inspection and Quality Department intervention.

As experience builds up the level of detail required might be reduced. For example, as measured historical data becomes available for complete activities such as preform manufacture or mould preparation these figures can be used, rather than using the individual sub-actions approach. In addition to this build-up of labour hours through individual actions some consideration needs to be given to the areas of working efficiency and learning curve effects. If historical data are used these effects are essentially accounted for, even if the precise efficiency figures might be hard to detect. If estimated figures are used some allowance for working efficiency will be needed.

For the tracking of the effects of design changes and design options on production costs the fully built-up labour costs approach is often adequate if used with care, unless there is the intent to replace large elements of the total labour input with automated processes. If labour is replaced with capital the same overheads burden per manhour cannot be assumed and a more sensitive costing approach is required.

12.3 PRODUCTION COSTING

Adequate production costing is in essence an extension of the outline costing approach used above, but with fully burdened manhour rates replaced with real labour rates and the fixed and variable overheads separately identified. Allowances should also be made for changes in overhead structure as manufacturing numbers change.

When production plans are drawn up, the number of actions or interventions required during the production cycle (from Commercial, Materiel, Engineering, Planning and Quality etc.) can be identified. Labour hours and other costs can then be allocated to these. These can then form part of the variable 'overhead' costs associated with the production line. Additional variable costs might be any disposable materials not included in delivered hardware and power consumption etc.

In addition there will be fixed costs on the production facility such as rent, general heating, tooling and production equipment, R&D, recovery of development funding, general marketing and management etc. The apportionment of these costs to a specific product line has to be decided so that estimates of total project costs and sales income can be generated with respect to sales volume, and thus sales prices can be established. It is usual to think of such costs as being apportioned as the proportion of total factory sales that arise from the particular product line. This is fine if the output is invariant. If this is not the case it makes more sense to treat these costs as a fixed overhead that the component-specific line is expected to bear, irrespective of output.

Before such practices as the ones outlined above can be used in any plant, the cost structure for the whole plant must be known, as it is not realistically possible to operate several different costing systems within one factory. This may be a burden for plants that currently have a narrow range of products and processes, where the simple built-up labour rate approach can be used without too much inaccuracy. On the other hand it will be very difficult to obtain adequately accurate costings for RTM work without taking some account of the changes in the overheads mix compared to other processes.

In an ideal world, the overheads structure of the company would be incorporated into a computer-based costing model so that what-if

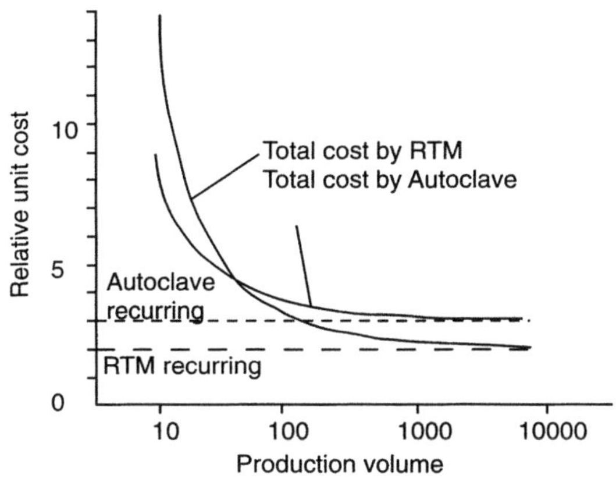

Figure 12.1 Comparative costs for simple components made by RTM and autoclave moulding.

questions can be readily answered, for example as to the effect on total production costs of introducing automation to reduce labour contents.

It will be noted that cost estimation is assumed here to be an activity that commences at the very beginning of the design process and is continually refined and kept up to date during design, prototyping and production readiness phases. Concurrent design and development practices are very important in the development of composite components. The great flexibility of RTM in terms of parts integration, net-shape moulding etc. makes it especially important that the costs of these options are taken into account properly in the design and development process.

Some effort has been put into cost comparisons between RTM and other advanced composites manufacturing processes. Krowleski and Gutowski (1989)[1] suggests lower costs for RTM at all production volumes; compared to autoclave moulding. This is rather surprising in view of the higher tooling costs for RTM: My calculations (comparing autoclave and RTM costs) indicate higher costs by RTM for flat panels at very low production volumes where tooling and other non-recurring costs are dominating (see Figure 12.1).

The break-even volume would be expected to be lower for more complex parts, perhaps as low as one off where the ability to produce high levels of parts integration can reduce the total tool and jig bill to below the levels for autoclave moulding several parts and bonding them together. Comparisons with other processes could only be made on a case-by-case basis and would very much depend on the detailed component as well as the process requirements. For example, lower costs would be expected for contact moulding at small volumes, unless tight tolerances would require extensive post-moulding machining of the contact moulded parts.

REFERENCE

1. Krowleski, S. and Gutowski, T. (1989) Economic comparison of advanced composite fabrication technologies, *Proc. 34th SAMPE Symposium*, 329–40.

13 Quality control/assurance

The word quality can be used in two distinct senses: firstly as some measure of perfection and secondly as a system for demonstrating that components have been made in an acceptable manner and can be documented to be acceptable. Questions of perfection in the normally accepted sense are out of place in a discussion of Quality systems, as the only useful definition is whether a component meets specifications or not. A component that only just meets specifications is deemed to have the same quality as a part of perfect exhibition standard. The design and development process has the greatest effect on quality defined in this way, especially with regard to the development of process windows that have the maximum flexibility with respect to conditions leading to stable product quality. These sorts of issues are considered elsewhere in this book and will not be repeated here.

This chapter is concerned solely with the sorts of controls that should be imposed on the process in terms of inspection points, and documentation control etc.

The function of a quality system is to demonstrate to the customer or regulatory authority that the products delivered are built to the required standard and are thus fit for purpose. The level of detail required to do this will depend largely on the function of the part, and critically on the consequences of product failure.

The elements of an effective Quality system are common to any manufacturing operations, and thus will not be dealt with here. As far as is practical only those aspects of Quality that are of specific relevance to RTM produced components will be considered here. It will be assumed that the components produced have at least some adverse consequences of failure and that a formal quality system is appropriate. While it is possible to argue that a formal system is not strictly necessary in small organizations outside aerospace, the increasing emphasis on product liability makes the adoption of some formal framework at the very least prudent at any scale of operations. In the worst case, if a failure of a product leads to loss or injury, any company that is not able to document that the part left the factory in accordance with its specifications is wide open to legal assault with no basis on which to construct a defence.

13.1 DOCUMENTATION REQUIREMENTS

There will be a need for some top-level control document that covers the applications of RTM processing within any particular organization. This might be termed a process specification or be given some other, similar title. This document will contain the basic information that all parties involved in the design and manufacture of RTM components need to share – at all levels from the design and development function to the shop floor. All the other documents relevant to the processing should be referred back to this one document so that it remains a complete reference. A specimen document to cover this sort of requirement is given at the end of the chapter. This specimen is only intended as a guide to the sort of documentation that might be required as each product and company has their own detailed requirements.

Any materials used in the process that form part of delivered hardware will require purchase specifications and, in many cases, specifications for acceptance testing of incoming materials. Although not strictly a Quality issue, Health and Safety data must also be available for any materials used in production, irrespective of whether they form part of deliverable products.

For every specific product to be made by RTM some sort of manufacturing instructions document will be required in addition to component drawings. This document details the process to be used for that specific component; what materials, cutting patterns, preform and other tools are used; material handling, layup and process conditions; finishing operations etc. For more critical applications requiring a high degree of traceability a record sheet giving operator and inspector sign offs will be needed in addition to the manufacturing instructions. An inspection document which gives details of inspections to be carried out and acceptance criteria for any minor defect types, will also form part of the product specific documentation.

Operating and safety instruction documents will also be needed for each item of machinery used in the production cycle, even if the machinery is normal commercial equipment that is supplied with operating manuals. Safety requirements can be taken separately from the Quality system, but integrating the two systems can have advantages.

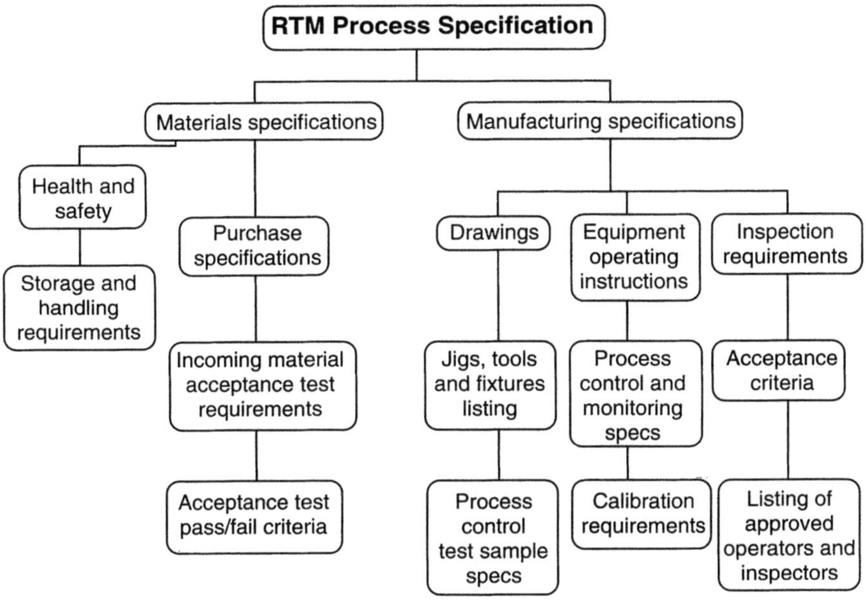

Figure 13.1 Quality control documentation usually follows this sort of hierarchial system.

For example, maintaining a listing of which operators are qualified to operate particular equipments will be needed for both Safety and Quality systems and it makes obvious sense to maintain one list rather than two.

In principle, all these documentary requirements could be merged into a single document. In practice, each company should have its own Quality Assurance procedures and the necessary documentation should be incorporated into the company system. It is perhaps most usual to have a hierarchical system of documentation such as that shown in Figure 13.1 fitting in within the overall Company Quality, Safety and Operational Systems.

The level of documentation required will be very high when critical components are being produced and much lower for most other parts. In aerospace, extensive documentation systems are generally required for all parts, the loading simply increasing for the most critical components. Even if the products that are made have very low consequences of failure, it is essential that some basic documentation is in place. As an absolute minimum this must cover Health and Safety aspects of the process. Higher levels of documentation and control are strongly recommended as these provide a focus for best practice and assist in tracking down problems in production. It is also recommended that documents should be written in such a way that they can be used in operator and inspector training, as well as to control the process.

13.2 PROCESS CONTROL AND PROCESS MONITORING

For any manufacturing process, there is a need for some measure of process control; and for the monitoring of the effectiveness and accuracy of this control. For RTM it is suggested that control is required at the following steps in the process.

13.2.1 Incoming materials

1. Check that materials are as called up in the appropriate specification. If life limited materials are used check that they are within their use date.
2. For bound reinforcements to be used in preforms, check which side is bound (if materials bound on only one side are used) and that binder application is uniform.
3. Check that reinforcements to be used are free of manufacturing defects, or to be more accurate that the frequency of manufacturing defects is within specification limits.

13.2.2 Reinforcement handling and preforming

1. Check that the cutting templates or cutting tools called up in manufacturing instructions are available and undamaged.
2. Check that stacks of reinforcements are of the correct material, orientation and ply count, and are free of wrinkles and other defects such as misalignment beyond specification limits.
3. Check that any reinforcement stack processing equipment is operating at the required temperature and pressure cycle.
4. Check that bonded preform precursor reinforcement stacks are fully bonded, and do not show evidence of overheating or other defects.
5. Check that preform blanks cut from precursor stacks are of the specified geometry and are free from defects.
6. Check that preform rig is operating to specification requirements, with respect to temperature, pressure and tool alignment, and that safety guards or other operator protection systems are operational.
7. Check that the correct heat-up, pressure and cooling cycle is used.
8. Check that finished preforms are fully bonded, do not show evidence of overheating and are free of process induced defects such as wrinkles, folds, gross fibre misalignment, tears etc.
9. Check that preform geometry and dimensions fall within the required tolerances.
10. Check that preform storage conditions are as stipulated in specifications.

11. Check that preform kits for production are complete and that any required quality or production paperwork is included with each kit or batch of kits.

Similar sets of process controls can be stipulated for other preforming approaches such as stitching, the process controls will be specific to the preforming style used and have to be generated for the particular system to be used.

13.2.3 Preparation of resins and resin injection equipment

(A) For resins that are to be manually mixed and injected via pressure pot equipment

1. Check that the resin constituents are as required and are within life.
2. Check that resin weighing equipment is of the required degree of accuracy and calibration.
3. Check that the weights of resin, hardener, accelerator etc. are within specified tolerance limits.
4. Check that any mixing equipment used is operating as required and is clean and free of contamination.
5. Check that all the necessary components of the resin are included in the mixing vessel and that the mixing conditions are as required, e.g. time, temperature, stirrer rpm and blade type etc.
6. Check that the mixed resin is homogeneous and free of contaminants.
7. If degassing is specified, check that the equipment is functioning properly and that the degassing conditions applied are as specified.
8. Check that the injector is functioning correctly, without any leakage or constrictions in output lines and within the specified temperature limits.
9. If factory compressed air is used to drive the resin from the pressure pot, check that the air quality is acceptable (i.e. at the required pressure and free of oil and water).

(B) For resins that are to be injected via mix and meter machines

1. Check that the machinery is charged with the resin constituents that are called up in the manufacturing instructions.
2. Check that the machinery is operating within specification, without evidence of backlash, wear in linkages or other indications of malfunctioning; and that the set mix ratio is as required.
3. Occasionally check that the mix ratio set on the machine is identical to that being delivered.

4. If the resin components are being stored in heated tanks, check that the residence time in the tanks is within acceptable limits, advice may be required from resin suppliers to establish these limits.
5. Check that the resin supply from the machine is properly mixed and within the specified temperature range.
6. Check that the machine is supplying resin at the required pressure/flow rate.

13.2.4 Preparation of mould tools for loading of reinforcement

1. Check that the correct mould tool is being used.
2. Check that the tool is fit for use. That is to say is clean and free of any damage, that ejectors and any other moving parts function properly, that seals are in place, and that the tool's release coating is in good condition.
3. Check that any required control equipment such as thermocouples, resin sensors and cure monitors are all properly installed, undamaged, in working condition and that any necessary calibration stamps are current.
4. Check that any in-mould heaters are in good working order.

13.2.5 Loading of mould with preforms or other reinforcements

1. Check that the mould loading environment is as required, e.g. temperature, lighting, humidity and cleanliness.
2. Check that the required reinforcement/preform kit is complete and correct.
3. Check that preforms are loaded in the order laid down in the manufacturing instructions, without the use of excessive force and without damage.
4. Check that no fibres are laying across any mould seals or trapped between any loose blocks in the tool. For pinch-off tools check that the necessary amount of fibre is present in the pinch region.
5. Check that the mould closes to the required dimensions and that the force required to shut the tool is within the required limits.
6. Check that the mould is properly clamped shut.
7. Check that the in-mould vacuum is at the required level (if vacuum is used) for the correct length of time; check that the leak-back rate is within specification.

13.2.6 Resin injection

1. Check that the injector is correctly coupled.
2. Check that both mould and resin are within specified temperature limits.

3. Check that the injection pressure/flow rate is within specified limits.
4. Check that the time of injection is within specified limits.
5. Check that all out gates have been showing a resin trace for the required time prior to stopping injection.
6. Check that the resin injector is demounted and completely flushed of mixed resin (if required).
7. Check that the mould is sealed according to the requirements of the manufacturing instructions.

13.2.7 Resin cure stage

1. Check that the time/temperature cycle applied is within specified limits.
2. Check that the mould and moulding are at the specified temperature for demoulding.

13.2.8 Demoulding

1. Check that the correct demoulding equipment is available and correctly functioning, and is used in accordance with the manufacturing instructions.
2. Check that any in-mould testpieces are properly identified and routed for test with the associated paperwork.
3. Check that the correct mould cleaning procedures are followed.

13.2.9 Post-moulding inspection

1. Check that the general moulding quality is within the acceptance criteria laid down for that particular component.
2. Check that the dimensions of the moulding are within tolerance.
3. Check that the weight of the moulding is within tolerance.
4. Check that all paperwork conforms to requirements, including process records, batch numbers, signatures etc.
5. Check that any measurements made on in-process test-pieces meet the requirements.

In most instances all these checks would be incorporated into the general operations of the production line. For critical components it would be expected that documentary evidence be provided to demonstrate that each of the checks had been carried out and that all materials, processes, equipment and moulded components fully meet all elements of their specifications. In addition any equipment used for the checks must itself be regularly calibrated against traceable standards. The assurance that the checks have been carried out would come from the use of a record sheet to be signed off by operators and inspectors as the process proceeds. If in-service failures of components ever led to any sort of legal action it would be necessary to be able to prove that components were adequate at the factory gate. In the absence of good records this would be an impossibility.

13.3 SPECIMEN DOCUMENTS

These are not intended to be complete or adequate for production purposes as such documents must be written within the framework of whatever Quality Assurance system is in use in individual plants. The specimen documents are only intended to indicate the sort of approach that may be useful.

13.3.1. RTM PROCESS SPECIFICATION.

Sheet 1 of 7

COMPOSITES CONSULTANCY LIMITED

PROCESS SPECIFICATION NUMBER: PSRTM1

TITLE: RESIN TRANSFER MOULDING OF PRODUCTION COMPONENTS.

ISSUE AMENDMENT

This space to be used to record amendments and changes to the specification.

Some system needs to be in force to ensure that changes can only be made by authorized persons.

Issue No	Date of issue	Author	Amended by	Production engineering	Change control
1					

1. SCOPE OF SPECIFICATION

Sheet 2 of 7

This document describes the general procedures to be used for the moulding of products by Resin Transfer Moulding (RTM).

It covers the manufacture of reinforcement preforms and their impregnation with suitable resins systems.

2. MATERIALS OF CONSTRUCTION

All materials must be purchased to the appropriate specification. Materials forming part of deliverable products must be purchased to a Certificate of Conformity to that specification.

2.1. Fibre reinforcements, resin systems and associated process requirements and tolerances shall be detailed in product specific specifications and procedures. Only materials that have been approved for use in RTM processing shall be used. A listing of approved materials is maintained by the design department.

2.2. Preform Binders shall be of powder, fibril veil or emulsion types.

2.2.1. Powder binders to Materials Specification: MSRTM1

2.2.2. Fibril veil binders to Materials Specification: MSRTM2

2.2.3. Emulsion binders to Material Specification: MSRTM3

3. ANCILLARY MATERIALS, NOT FORMING PART OF DELIVERED PRODUCTS

Number	*Type*	*Approved materials*	*Notes*
3.1	Paste release agent		Use on new tools
3.2	Liquid release agents		Use on operating tools
3.3	Sealing material	Rubber ‘O’ ring	For sealing moulds
3.4	Adhesive tape	Flashbreaker	To mask surfaces of tools from cured resin contamination.
3.5	Air bleeder fabric		To assist mould venting where needed.
3.6	Release film		For use on flat moulds as an alternative to release agents
3.7	Cleaning solvent		Use in preparing mould surfaces to receive release agent.

Issue 1. PSRTM1

Sheet 3 of 7

4. MOULD TOOLS

4.1. All tools shall be designed and manufactured to appropriate standards to give an adequate lifetime consistent with the production quantity required. Tooling materials shall be chosen consistent with this aim and with the geometrical and other tolerances imposed on the part.
4.2. All new or reconditioned tools shall be given a full release agent coating prior to use. Surfaces shall be solvent washed, a coat of sealant paste shall be applied and polished in accordance with the manufacturer's instructions. A minimum of three coats of an appropriate release agent shall be applied, with air drying between applications, followed by curing to the manufacturer's recommendations. Any changes in resin type used in any mould shall be a reason to check the compatibility of the new resin with the release agent on the tool, only release agent/resin combinations that have demonstrated compatibility and are incorporated into the listing in this document shall be used.
4.3. Tools in production use shall be examined regularly for damage to the mould face or sealing surfaces. Any damage noted such as may lead to changes in geometry, poor sealing or difficulties in mould closure or ejection shall be a reason for remedial action. Damages to mould surfaces shall not be corrected on the shop floor without authority. Any repaired mould requires an Inspection sign-off prior to release to the shop floor and an application of the necessary release coating before use.
4.4. Prior to any use the tools must be examined for any resin contamination or minor damages to release agents. All resin contamination or release agent buildup must be removed prior to use. Only approved non-scratching scrapers are permitted for this duty, plastic, brass or composite scrapers may be acceptable. Advice on the suitability of scrapers and other tools used to prepare mould surfaces can be obtained from the Tooling Department.

5. RESIN INJECTION MACHINERY

5.1. The primary function of the resin injection machinery is to provide a supply of pressurized resin to the resin in-gate of the mould. In addition, the machinery may mix the resin/hardener combination.
5.2. Pressure pot machines are simple and robust, premixing of resin is required and maximum pressure capability is low. Acceptable machinery of this type is supplied by the following companies...............
5.3. Mix and meter machines are more complex and it may be more difficult to ensure that resin is properly metered, mixed and degassed. Shot size and pressure capability are high. Acceptable machinery of this type is supplied by the following companies...............
5.4. For any production job, one or other type of machinery will be chosen prior to qualification of the production line and moulded components.

5.5. No changes in type of machinery or operating parameters are permitted without the approval of the product design authority. Requalification of the production line or moulded components may be required. The extent of requalification required will be determined by the Quality department and product design authority.

6. REINFORCEMENT HANDLING AND PREPARATION OF PREFORM BLANKS

6.1. All handling of reinforcements, preforms and mould loading shall be carried out to the following set of environmental conditions.
6.1.1. Shop temperature 15–30 °C.
6.1.2. Relative humidity max 75%
6.1.3. Floors shall be sealed and regularly cleaned with vacuum cleaners, sweeping is not permitted.
6.1.4. Internal combustion engines are not permitted in the working area.
6.1.5. Smoking, eating and drinking are not permitted in the working area.
6.1.6. All materials, tools, moulds and other equipment taken into the area shall be free of dirt, grease, oil etc.
6.1.7. Temperature and humidity may require monitoring and or recording.
6.1.8. Reinforcements and preforms shall at no time come into contact with ungloved hands. Clean, lint free gloves shall be worn by all operators coming into contact with reinforcements. Release agents and barrier creams shall not be permitted to come into contact with reinforcements.
6.2. For components of some complexity, methods need to be used to hold the reinforcement in the correct shape prior to mould closure and resin injection. The name given to these techniques is preforming.
6.3. Preforming can utilize methods analogous to those used with prepreg or be produced by press forming. In either case the materials to be used must first be prepared by the addition of some binder material. Other methods of preforming such as stitching may be permissible, details are to be found in Component Specific Manufacturing Instructions.
6.4. Reinforcements may be purchased with binder already applied or binder may be applied as part of the processing. Binder application will depend on the type of binder used and shall be controlled by Process Control sheet number PCRTM2. The quality of reinforcements with binder applied shall be controlled by Materials Specification numbers MSRTM1, MSRTM2 or MSRTM3.

Sheet 5 of 7

7. 7. PREFORMING

7.1. Preforms may be made from single plies of reinforcement, but are more usually made from multi-ply stacks. The number of plies, their orientation and materials shall be in accordance with drawing requirements. Where single-sided binder application is used care shall be taken that the stack is made with the bound side of the reinforcement in contact with the unbound side of the adjacent ply.

7.2. Ply stacks shall be heated to a temperature such that the binder is softened, but neither the fibres nor any surface treatments are damaged, and that the binder is not oversoftened. When the correct temperature range has been reached the ply stack shall be pressed in cold tooling and held until it has cooled sufficiently to be handleable. Vacuum bag pressure is generally adequate to produce well-bound ply stacks.

7.3. To generate shaped preforms some form of tooling is required. Tooling materials may be wood, plastic, GRP, or metal. Purpose-built preforming rigs are required for accurate production preforming.

7.4. Preform tools should be checked for wear, damage and correct alignment prior to any use.

7.5. Preform blanks shall be cut from ply stacks, without damage to the reinforcements. The developed shape and ply orientation of the blank shall conform to drawing requirements. Blanks shall be heated and presented to the preform tool in such a way that the correct orientation of the blank is guaranteed. Pressure shall be maintained on the formed blanks until they have cooled enough to be handleable. Cooling time shall be established experimentally, but for most preforms 30 seconds should be adequate.

7.6. The preformed reinforcement shall be inspected for damage due to the preforming process. Acceptable conditions for each preform in each component shall be included in part-specific acceptance criteria.

7.7. Preforms can be trimmed to size with knives, scissors, blanking tools etc. The procedures to be followed for any particular product shall be laid down in Component Specific Manufacturing Instructions.

7.8. Completed preforms should be stored in a cool dry place, protected from mechanical damage and any sources of contamination such as oil, water, dust etc.

7.9. Any inserts that are to form part of the complete kit of 'preforms' for a component, e.g. metal inserts, foam cores, should receive the recommended surface preparation as laid down in the relevant Process Specification.

8. MOULD LOADING

8.1. Check mould tool for cleanliness and state of the release agent.
8.2. Check all moving parts and seals.
8.3. Following the procedure and order of loading laid down in the Component Specific Manufacturing Instructions, insert each preform into place, taking care not to distort preforms or move the position of preforms already within the tool. Aids to loading may be used as laid down in Component Specific Manufacturing Instructions.
8.4. Check the completed layup for material that will prevent the mould from closing properly or bridge the mould seals.
8.5. Place the top of the mould in position and close the mould carefully to stops. If the mould fails to close properly or requires excessive force in closure it must be carefully reopened, the cause of the problem must be identified and rectified. Damage to tools and the rejection of moulded parts can be caused by failure to rectify preform faults.

9. RESIN PREPARATION

9.1. The range of resins used, and the differences in the handling and treatment of these depending on what sort of resin injection machinery is used, make it impossible to cover all possibilities in this document. Resin system handling and preparation will therefore be covered in individual Component Specific Manufacturing Instructions.

10. RESIN INJECTION AND RESIN CURING

10.1. Check that the mould temperature is within the tolerance as required for the specific tool.
10.2. If required, apply vacuum to the tool, check that the tool reaches the required level of vacuum, and that the tool is free of vacuum leaks outside of acceptable levels, as laid down in Component Specific Manufacturing Instructions.
10.3. Connect up resin lines, if these do not form part of a manifold with the vacuum/outlet lines.
10.4. Inject resin within the temperature limits and at the pressure or flow rate specified in the Component Specific Manufacturing Instructions. When all the resin outlets are showing clear resin without entrained air switch off the vacuum pump (if this has been used). Cease injection of resin, disconnect resin lines and plug mould.

Sheet 7 of 7

10.5. Ensure that the mould temperature reaches and is stabilized within the required temperature range for the required time, as laid down in Component Specific Manufacturing Instructions.

11. DEMOULDING

11.1. Mouldings are usually removed from the tool at room temperature. If it can be verified that warm demoulding causes no harm to the moulding or tool, then demoulding can be carried out at elevated temperature. Maximum demoulding temperatures are to be found in the relevant Component Specific Manufacturing Instructions.
11.2. If hot demoulding is carried out components shall be held in suitable fixtures during cooling to room temperature to prevent distortion.
11.3. Demoulding shall only be carried out with those ejectors that are built into the tool, or with external ejector mechanisms specifically designed for the tool. No tools other than those specifically designed for use as ejectors are permitted to be used.

13.3.2. COMPONENT SPECIFIC MANUFACTURING INSTRUCTIONS

Sheet 1 of 4

COMPONENT DESCRIPTION: Reflector dish.
COMPONENT NUMBER: CCL/RTM101

Related documents:-
RTM process specification no PSRTM1
Materials specifications nos MSRTM1. MSRTM8. MSRTM13.
Drawing numbers CCLDRTM101,
Preform drawing numbers CCLDRTM101/P1 & 2
Gauges number CCLRTM101/G1 & 2
Preform tool numbers CCLDRTM101/PT1 & 2
Preform blank template numbers CCLDRTM101/1B1 & 2
Mould tool number CCLDRTM101/MT1
Ejector tool number CCLDRTM101/T1
Trim fixture number CCLDRTM101/T2
Inspection plan number CCL/RTM101IP1

ISSUE AMENDMENT

This space to be used to record amendments to the specification.

Some system needs to be in force to ensure that changes can only be made by authorized persons.

Issue No	**Date of issue**	**Author**	**Amended by**	**Production engineering**	**Change control**

1. PREPARATION OF REINFORCEMENTS

1.1. Withdraw 2 m of glass cloth code RTM1 from floor stock or stores, record batch number.
1.2. Inspect for flaws in cloth.
1.3. Using templates CCLDRTM101/1B1 & 2, cut four plies according to drawing CCLDRTM101/1 and four plies according to drawing CCLDRTM101/2.
1.4. Place ply 1 on layup bench, bound side uppermost.
1.5. Place plies 2–4 on top of ply 1, bound side lowermost. Ensure that ply directions are constant throughout (i.e. 4 plies at 0.90)
1.6. Place ply 5 on layup bench, bound side uppermost.
1.7. Place plies 6–8 on top of ply 5, bound side lowermost. Ensure that ply directions are constant throughout (i.e. 4 plies at 0.90)
1.8. Place a sheet of release paper over the two ply stacks. With an electric iron set to 130 °C, carefully press the ply stacks to bond together the reinforcements using light pressure. Check that the plies are well bonded together.

2. PREFORMING OF REINFORCEMENT STACKS

2.1. Ensure that preform rig has reached operational temperature.
2.2. Mount preform tool CCLDRTM101/PT1 in the press.
2.3. Place ply stack 1–4 under heaters for 1 minute, maximum permitted surface temperature is 130°C, as measured by thermocouple at the surface of the stack.
2.4. Transfer stack to press station and close press to form preform 1.
2.5. Remove preform 1 when cool and check for wrinkles, folds and other defects. Trim to size to drawing CCLDRTM101/P1. Place preform 1 in kitting tray.
2.6. Mount preform tool CCLDRTM101/PT2 in the press.
2.7. Place ply stack 5–8 under heaters for 1 minute, as before.
2.8. Transfer stack to press station and close press to form preform 2.
2.9. Remove preform 2 when cool and check for wrinkles, folds and other defects. Trim to size to drawing CCLDRTM101/P2. Place preform 2 in kitting tray.

2.10. At production convenience the number of preforms 1 and 2 in any batch can be varied to obtain the most effective working practices.

3. INSERT MANUFACTURE

3.1. Foam insert to drawing CCLDRTM101/3 is required between preforms 1 and 2.
3.2. Foam insert blocks are to be machined to shape from foam material RTM8. Finished foam blocks are to be inspected using gauges CCLGRTM101 & 2

4. MOULD ASSEMBLY

4.1. Obtain mould number CCLDRTM101/MT1 Check for condition and apply mould release as required.
4.2. Insert preform 2 into the cavity in the mould tool.
4.3. Insert foam insert into the cavity in preform 2 in the mould tool.
4.4. Insert preform 1 into the mould tool.
4.5. Check that preforms are correctly positioned and will not prevent mould closure.
4.6. Place mould tool upper half into position, close tool and check that tool has closed to stops as required.
4.7. Apply vacuum to the mould, required vacuum level is 10 mBar minimum, not leaking back to a level above 50 mBar after 1 minute with the vacuum pump isolated from the tool.
4.8. Stabilize tool temperature to 70 °C +/–5°C.

5. RESIN PREPARATION

5.1. This moulding requires 1 kg of mixed resin, using resin mix number RTM13
5.2. The resin components shall be weighed out to an accuracy of +/–1 gram.
5.3. The resin shall be mixed at ambient temperature using a mechanical stirrer with a speed of 200 rpm, until the mix is of uniform colour and shows no evidence of uneven mixing. Mixing time shall be not less than 2 minutes.
5.4. Resin mix shall be transferred to a heating cabinet to raise its temperature to 60 °C +/–5 °C. Mixed resin must be used within 30 minutes of reaching the correct temperature.

Sheet 4 of 4

5.5. Resin shall be degassed for 5 minutes at 10 mBar prior to loading into the injector.
5.6. Mixed, heated and degassed resin shall be loaded into the injector without entrapment of air and raised to the injection temperature of 60 °C +/–5°.

6. RESIN INJECTION

6.1. The resin injector shall be coupled to the mould/vacuum system and a pressure of 3 Bar applied to inject the resin into the mould.
6.2. The outgates of the tool shall be monitored to detect the presence of resin. When resin is streaming from all outgates and is free of entrained air bubbles the injection shall be terminated.
6.3. Turn off the vacuum pump, turn off the injection pressure, decouple the injection system from the tool and plug the tool cavity.

7. CURE

7.1. The tool shall be heated to a cure temperature of 100 °C +/–5°, at a rate of 5 °/min maximum 2 °/min minimum. The cure temperature shall be maintained for 60 +/–10 minutes. Thereafter the mould shall be cooled to below 60 °C prior to demould.

8. EJECTION OF COMPONENT

8.1. Tool number CCLDRTM101/T1 shall be used to extract the moulding from the tool.
8.2. Moulding shall be deflashed and trimmed to size in tool number CCLDRTM101/T2.
8.3. Moulding and associated paperwork shall be routed to Inspection.

9. MOULD CLEANING

9.1. Clean mould of any contamination with flash, release agent buildup etc.
9.2. Return tool to tool storage area.

Issue 1. CSMI/RTM101.

RECORD SHEET. CSMI/RTM101 (issue 1)

REFLECTOR DISH.

Materials.	Batch numbers.
Glass cloth type RTM1	________
Foam type RTM8	________
Resin part A RTM13/1	________
Resin part B RTM13/2	________
Resin part C RTM13/3	________

Action	Operator	Inspector
1.1	______	
1.2	______	
1.3	______	
1.4	______	
1.5	______	
1.6	______	
1.7	______	
1.8	______	
2.1	______	
2.2	______	
2.3	______	
2.4	______	
2.5	______	
2.6	______	
2.7	______	
2.8	______	
2.9	______	
3.2	______	______
4.1	______	
4.2	______	
4.3	______	
4.4	______	
4.5	______	
4.6	______	
4.7	______	
	Vac after 1 min =	
4.8	______	
	Tool temp =	

Action	Operator	Inspector
5.2	______	
	Weight resin part A =	
	Weight resin part B =	
	Weight resin part C =	
5.3	______	
5.4	______	
	Time at end of resin mix =	
	Time at holding temperature =	
	Time at use of resin =	
5.5	______	
5.6	______	
6.1	______	
6.2	______	
6.3	______	
7.1	______	______
8.1	______	
8.2	______	
8.3	______	______
9.1	______	
9.2	______	

Part inspection.

Dimensions OK.	______
Acceptance criteria OK.	______
Part not acceptable.	______
Reason for rejection.	______
Corrective action.	______

Case study 14

14.1 INTRODUCTION

It is not often possible, in the publicly available literature on RTM, to discern details of the construction, what if any preforming techniques are used, reasons for the selection of particular mould design details etc. This is doubtless done for sound commercial reasons, but can lead to limitations in the utility of the information. For this reason a wholly imaginary component will be considered here, which incorporates some of the features that have been reported in the literature.

For the purposes of this case study it is assumed that the design has been subjected to a stress analysis and will have an adequate mechanical

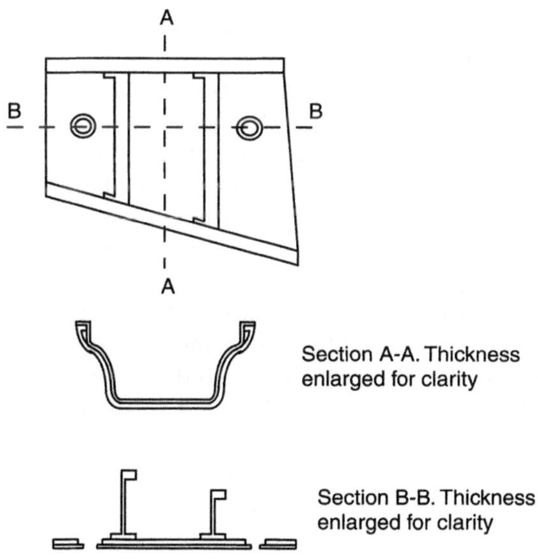

Figure 14.1 Sketch of case study component.

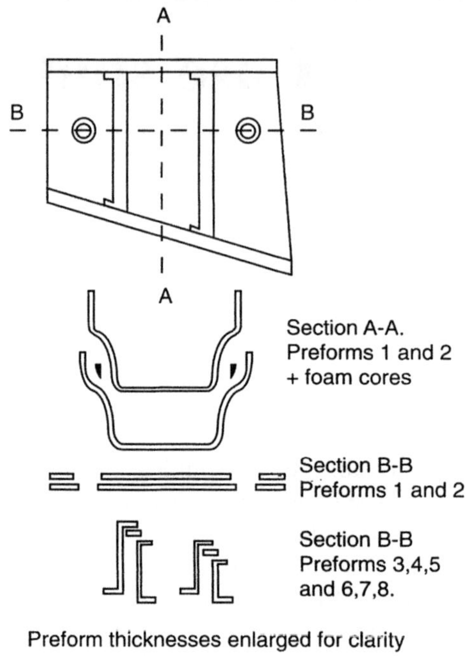

Figure 14.2 Exploded view of preforms in component.

performance in its use environment. The component to be considered is shown in Figure 14.1.

It consists of an outer shell with two sets of cross ribs. Each edge of the part is to be net-shape moulded and the long edges incorporate a foam filled edge member for attachment to other structure by bonding, two holes are to be moulded in to carry bearing housings. The component shows undercuts, ribs and moulded in holes, and is net-shape moulded.

Eight major preforms are used, plus two foam cores: two outer shell preforms, and six preforms that comprise the two internal cross ribs. The six rib preforms could be consolidated into two complete rib preforms, leaving four preform assemblies to be inserted into the tool.

See Figure 14.2 for the preform breakdown. It is assumed here that a heat-softened powder binder will be used to rigidize the preforms.

14.2 PREFORM DESIGN AND TOOLING

14.2.1 Outer shell preforms

These are the simplest preforms as the shaping is essentially single curvature. The layup for this shell is, therefore, not constrained by considerations of any in-plane fibre direction rearrangements that would be required to generate double curvature. The two preforms would best be produced on individual tools. First the preform blanks have to be prepared and the holes for the bearing housing seats cut out in accurate register to each other. The simplest way to achieve this would be to mount two knife cutters on a single board and use a roller press to blank out the holes. Alternatively the holes could be hand-cut using a metal template. A master plug model can be readily made for such a simple shape and should be adequately accurate for preform tooling. From this master model a shell is cast off or laminated in GRP and mounted onto a suitable press board. This shell is then waxed to the required preform thickness using sheet wax, and a plug is cast off to form the inner part of the preform tool. Into this plug two insert blocks will be fixed to keep the bearing housing seat cutouts in the proper alignment. The two parts can then be mounted in a suitable preform press ready for use. This process can then be repeated for the other outer shell preform tool, using the plug from the first preform tool as a master. Alternatively the preform tools could be made without insert blocks and the bearing housing seat cutouts could be cut into the finished preforms. Both approaches have advantages, precutting the holes gives a self-aligning feature to the preform so that the trimming allowance can be minimized, but any movement of the blank during preforming could result in misalignment of the holes and a scrap preform.

If the design had required areas of variable thickness, these would have had to be made as separate preforms. For example, local reductions in ply count in low-stress areas in order to save weight and materials costs, would lead to increases in labour costs and a higher preform tooling bill for the preform manufacture. This is in contrast to the case for hand layup of single plies of prepreg where such changes have a smaller direct influence on manufacturing costs; although even in this case increased complexity does tend to lead to increased costs.

In principle, for such simple shapes as discussed here mechanical preforming would not be required and the reinforcements could be loaded directly into the tool. On the other hand this would tend to increase layup labour and extend the cycle time for the moulding. In addition, when reinforcements are preformed the reinforcement bulk factor is usually reduced, which helps to reduce problems with reinforcement movement during mould loading.

14.2.2 Cross rib preforms

The shape of these preforms has double curvature, therefore, a layup of +/–45° to the cross direction is preferred to 0\90 as this layup would have a tendency to wrinkling in the corners. It can be seen from Figure 14.1 that, due to the taper in the part, one of the preforms in each rib will have to be slightly distorted in mould loading. This would not be expected to be a problem with typical, semi-rigid, preforms.

The inner plug for the outer shell preform has the required shape for the edges of the cross ribs and can be used to generate these preform tools. Firstly, a shell is cast off the outer shell preform plug. This now has the correct geometry for the outside of the cross-ribs. As noted above, one side of the rib has an effective undercut with respect to the direction that the preform tool would move in. This undercut would have to be relieved by building up the plug locally prior to casting off the shell. This shell can be modified by an internal cross board which forms the

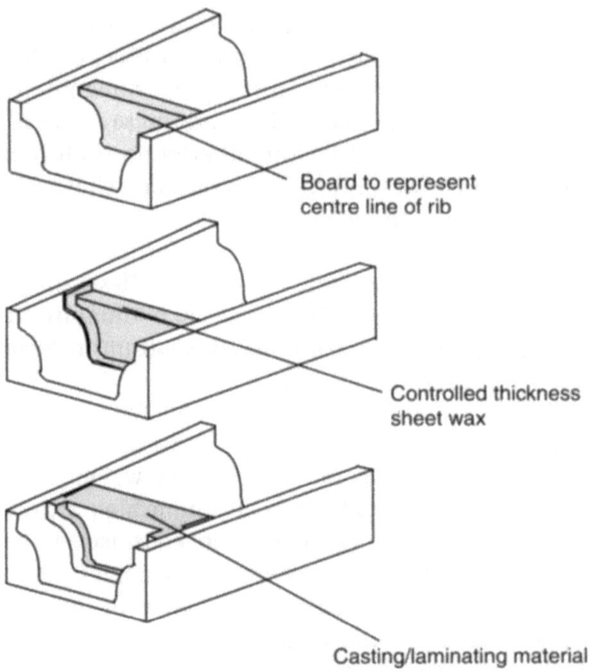

Figure 14.3 Stages in the production of cross rib preforms.
Stage 1. A splash is taken from the outer shell preform plug, and a board is located at the correct rib position.
Stage 2. Sheet wax is used to build up the required thickness of the preform.
Stage 3. Suitable laminating or casting material is used to build up sufficient thickness to form the preform tool.

shape of one side of one cross rib, and wax sheet can be used to build up the preform thickness against the cross board (see Figure 14.3).

The shell can now have a plug cast out of the cavity produced by the shell wall and waxed up boards. This plug now forms the inside of the cross rib preform; it can be used to cast or laminate an outer shell as before. This process can then be repeated for the other preforms. Alternatively a board can be made of the correct shape to replicate the internal cross-section of the part and this can be used to construct the preform tools. The last preforms required are two UD packers that are enclosed between the two pairs of cross rib preforms. No tooling is needed for these; they may simply be cut to size from suitable bound ply stacks.

The various preform tools could also be cast off the mould tool, rather than a master block. If production numbers were low it would not necessarily be economic to make a full set of preform tools, but the provision of single-sided tools for the cross ribs would greatly facilitate their production out of the tool.

Two approaches are possible to the further assembly of this set of eight preforms. They could be assembled directly into the mould, without any bonding between the preforms, or they could be bonded together in an assembly jig to give a full handleable preform. In the case here the first option would be preferred as the undercuts that are generated by the tops of the ribs are easiest to handle by assembling the rib preforms onto the appropriate tool blocks prior to mould assembly. Note that the preforms forming the outer shell need to be folded over to encapsulate the foam cores as part of the mould loading procedure, as shown in Figure 6.3.

14.3 MOULD TOOL DESIGN

The mould tool has to cope with a complex geometry and undercuts at the ribs. A simple tool of two parts is not possible in this case and a multipart tool is a requirement. A single shell can form the toolface in contact with the outside of the part. To clear the undercut between the two ribs, two mould parts are required. To clear the undercut at the second rib, only one mould part is required if it can slide in the axial direction. The mould assembly is completed by a top shell that is in close contact with all the other mating blocks. A single main mould seal is provided between the top and bottom shells.

A total of six mould parts might then be required as noted below:

1. inter rib mould block 1
2. inter rib mould block 2
3. open end mould block

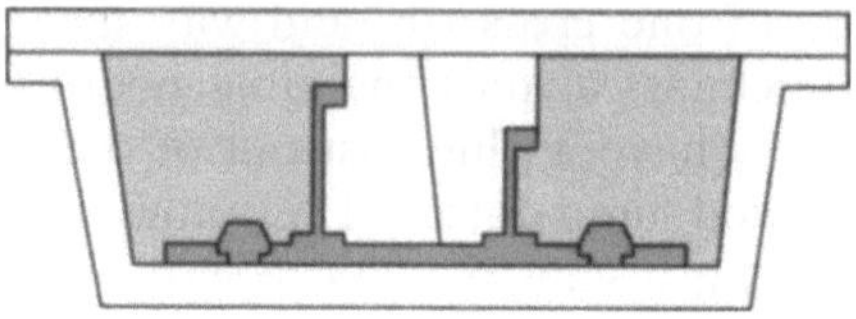

Figure 14.4 Baseline design of tool requires six separate parts to clear the undercut.

4. closed end mould block
5. bottom mould shell
6. top mould plate

These mould blocks are shown schematically in Figure 14.4.

Each mould block must have an ejector mechanism fitted to it, of one of the forms noted in the section on mould design. Some attention must now be paid to the siting of resin injection ports. These might conveniently be placed at the geometrical centre of each bay of the moulding, except in the one case where the bearing housing seat hole interferes with this choice. In this case it may be possible to use the seating hole as a ring gate, alternatively the resin port could be offset. Whichever solution is chosen it is necessary to check that the resin flow will not lead to dry spots in the moulding. This is perhaps most likely at the top of the rib. If resin flow simulation software is available this can be used to model the flow front development from different injection positions. Even in the absence of such software, simple, graphical techniques can be of benefit in predicting flow front shapes.

Figure 14.5 shows such a simple approach for this case. In Figure 14.5 the rib is assumed to be more permeable than the shell to ensure that the rib fills slightly ahead of the wall to prevent air entrapment. Figure 14.5 also shows how easy flow paths might distort the flow front shape and lead to air entrapment. Considerations such as this can assist in identifying suitable out gate positions and whether the use of evacuated tooling is a requirement.

Complex moulds such as that described above have been constructed and used with good results to produce high-quality components. Despite this, such moulds are obviously going to be relatively costly and may be more difficult to manufacture and operate in service.

Therefore, the question arises whether a simpler mould could be arrived at by modifications to the component geometry. In the case being considered here this would involve removal of the undercuts and the ensurance of an adequate draft angle to avoid any extraction problems.

This could be achieved if the solid crossribs were replaced with foam cored sandwich panels. Such crossribs could utilize additional transverse

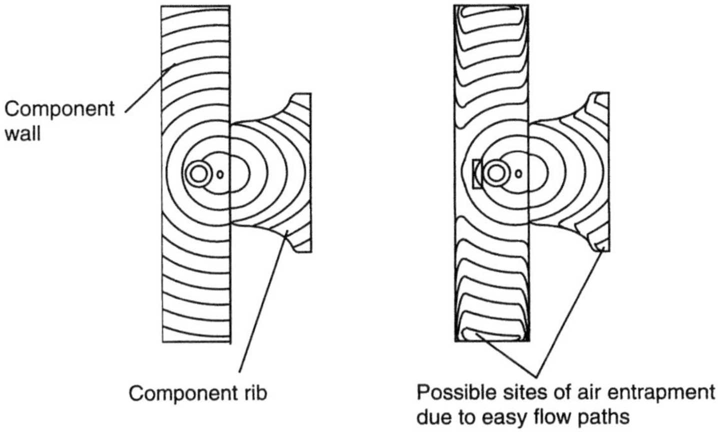

Figure 14.5 Simple estimation of the progress of the flow front can assist in the selection of in-gate positions. The general shape of the flow front is relatively easily mapped within 1 'cell'. Contours are not linked to time, but the flow in the rib and wall are linked, the permeability in the rib may be higher than in the wall to limit air entrapment. The influence of easy paths can also be estimated so that likely air entrapment positions can be identified.

UD material at the top of the ribs to ensure that stiffness was adequate. This approach would considerably simplify mould construction and use, but would increase the number of preforms required as two foam cores would also be required. The foam cores would have to be accurately made to control the overall permeability within the cross-ribs (see Figure 14.6).

This design would involve slightly more labour for preform preparation, but less for mould assembly and cleaning, and because the main mould elements are now fixed rather than loose, mould construction costs should be reduced and tolerances improved. Which design is to be preferred then depends on which activities have the greatest effects on costs and quality, and on the number of components to be made. In addition, unless the foam cores were accurately made they may create a more complex flow front and they also introduce the possibility for air entrapment in the outer shell below the foam in the cross ribs. The resin in-gate positions may need to be modified to account for the use of the foam cored ribs and the effect of the foam needs to be modelled. On the other hand the use of a two part tool reduces the possibilities for uncontrolled resin flow between the loose mould blocks. Such complex moulds would normally be used with vacuum assistance, reducing the potential problems in this area.

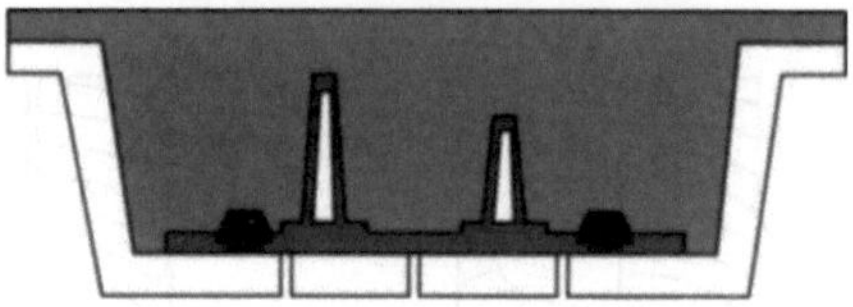

Figure 14.6 Revised design of tooling only requires two parts as undercuts are removed.

Overall the design with foam filled ribs would very often be the best option to follow.

Lastly such components could be manufactured by making the outer shell and cross ribs as separate parts and adhesively bonding them together. In this case other techniques than RTM could be used to produce the shell and ribs. While this appears to be a simpler option than the manufacture of one integrated component, tooling must be procured for each individual component and assembly/bonding jigs are required. If this option were to be followed the individual parts should be tooled to the adhesive bonding surface to ensure simplicity of assembly without excessive gap filling. The total tooling bill for the adhesively bonded option would be expected to be much higher than for the monolithic version; a minimum of three mould tools and one assembly jig are required. Labour costs would also be expected to be higher, as the total actions/activities listing would be much longer for the assembly case.

To conclude, a cost advantage would be expected in this case for integrating the cross ribs into the outer shell in a single RTM moulding. Once the decision is taken to make an integral moulding, modifications that simplify the tooling or layup required should be considered. While it may sometimes be easier to control process risks with the assembly approach, and thus reduce scrap/rework rates and costs, the higher baseline costs of this approach would still favour the integrated moulding unless its scrap rates were very high. The higher the level of integration chosen the more important is the establishment of adequate process controls and operator training.

Appendix A brief word about patents

Many patents exist in the RTM area. While this work has been prepared as far as possible from open literature sources it is possible that some of the recommendations made are covered, somewhere in the world, by patents. These patents fall into various areas such as resin chemistry and formulation, specific components made via the RTM route, RTM injection machinery, tooling, preforms and moulding. With regard to users of RTM technology rather than material or equipment suppliers those patents that apply to tooling, preforms and moulding will be the most important. A search of the available patent literature, covering six years, revealed more than 70 patent applications in these areas. This search was based on the use of key words such as resin transfer moulding, resin injection moulding, liquid injection moulding, and structural reaction injection moulding. The use of such, tightly focused, key words will inevitably lead to not all relevant patents being picked up. Indeed, I had two patents in my possession that were not found in the search precisely because their key words were not in the list used. An examination of the patent abstracts reveals that many applications seem to have been for concepts that are already in the prior art, in some cases for a considerable time. Examples include concepts such as the use of expanding foam cores, the use of air and solvents to flush systems and the use of resin channels in foam cores and other gating strategies such as the use of line gates to speed resin flow. In the latter case Farris *et al.* (1983)[1] discusses such techniques in a paper published five years before the relevant patent application. With regard to patents on internal gating such as might be formed by channels in foam cores, RTM theory is perfectly clear that any and all low Vf% zones will act as gates, whether they are produced intentionally or essentially accidentally, e.g. by chamfer angles on foam blocks or small gaps between tool blocks or preform blocks.

Overall, the patent situation seems rather muddy. Some simple concepts seem to have been patented more than once and the extent to which many of the patents that make broad claims were both novel and not obvious at the time of filing is at best questionable. A thorough review of the patent literature would be a gruelling task as a very wide search would be needed and a massive amount of irrelevant data would inevitably be generated. Reaching an opinion on the likelihood of specific patents being upheld if challenged would require detailed professional advice, which is far beyond the scope of this work.

Having said all of the above there do not appear to be any strong 'master patents' that limit what individuals can achieve within the generality of RTM technology. To end this brief statement it must, unfortunately, be repeated that while I have made various recommendations and suggestions in this work it cannot be assumed that they may be freely followed without any possibility of patent infringements.

REFERENCE

1. Farris, R., De la Mare, H., Overcashier, R. and Gottenberg, W. (1983) Structural parts from epoxy RIM using preplaced reinforcements, *Polym. Plast. Technol. Eng.* **21**(2), 129–57.

Index

MIX
Papier aus verantwortungsvollen Quellen
Paper from responsible sources
FSC® C105338

If you have any concerns about our products,
you can contact us on
ProductSafety@springernature.com

In case Publisher is established outside the EU,
the EU authorized representative is:
Springer Nature Customer Service Center GmbH
Europaplatz 3, 69115 Heidelberg, Germany

Printed by Libri Plureos GmbH
in Hamburg, Germany

SpringerBriefs in Economics

For further volumes:
http://www.springer.com/series/8876